Contents

Contents 1

Introduction 9

1 Logarithms and exponentials   11

    1.1 Main definitions and properties 11

       1.1.1 Power properties 11

       1.1.2 Logarithm definition and properties 14

    1.2 Exponential Functions 16

    1.3 Logarithmic Functions 20

2 Boolean algebra   25

    2.1 Statements and operators 25

    2.2 Truth Tables 26

       2.2.1 Proving the fairness of a statement 30

    2.3 Tautologies and contradictions 31

      2.3.1 What is really necessary? 34

  2.4 Quantifiers 35

3 Functions   37

  3.1 Basic definitions 37

  3.2 Further definitions 39

  3.3 Compound function (function of another function) 47

  3.4 Inverse functions 49

  3.5 Inverse of the elementary functions 51

      3.5.1 Power functions with odd exponent 51

      3.5.2 Power functions with even exponent 51

      3.5.3 Exponential functions 53

      3.5.4 Function $\sin x$ 54

      3.5.5 Function $\cos x$ 54

      3.5.6 Trigonometric tangent 55

      3.5.7 Particular functions 55

      3.5.8 How to write the algebraic expression of the inverse function . 56

  3.6 Absolute value (modulus) 57

  3.7 Integer part function (floor) 59

4 Binary Relations   62

	4.1	Relations properties	62
	4.2	Equivalence relations	64
		4.2.1 Properties of the equivalence sets	64
		4.2.2 An economic application	65
	4.3	Order theory	66
		4.3.1 An economic Application	67
	4.4	Suggested exercises	68
		4.4.1 Solved excercises	70

5 Numerical sets 75

	5.1	Natural numbers	75
	5.2	Integer numbers	78
	5.3	Rational numbers	80
		5.3.1 Is the rational set enough?	80
	5.4	Real numbers	82
	5.5	Finite and infinite sets	85

6 Topology of the real axis 87

	6.1	Basic definitions	87
	6.2	Classification of the points of a set	89
	6.3	Extremes of a set	92

7 Limits — 97

7.1 Universal definition of limit — 98
7.2 Subcases of the universal definition — 99

8 Theorems on limits — 114

9 Limits computation — 122

9.1 Partial Algebra — 122
9.2 Missing (Unlucky) Cases — 128
9.2.1 Algebraic cases — 128
9.2.2 Exponential cases — 131
9.3 Fundamental limits — 132
9.4 Asymptotic relations — 135
9.4.1 Sorting of the infinites — 137
9.5 Elimination (reduction) principle — 138
9.6 How to solve missing cases? — 139

10 Continuous functions — 142

10.1 Basic definitions — 142
10.2 Discontinuities — 144

11 Theorems on continuous functions. — 150

11.1 Basic definitions: 150

11.2 Theorems 151

11.3 Transformations of continuous functions 154

12 Derivatives 162

12.1 Definition of derivative 162

12.2 Derivatives of the elementary functions 164

12.3 Computation Rules 166

12.4 Points where a continuous function is not smooth 169

12.5 Theorems on smooth functions 173

    12.5.1 Some notes on De L'Hopital rule 179

12.6 Differential 180

    12.6.1 Equation of the tangent line 182

12.7 Series expansion 183

    12.7.1 Taylor's polynomial 183

    12.7.2 McLaurin's polynomial 187

12.8 Rules to determine greatest and smallest of a function 187

12.9 Concave an convex functions 189

13 Graph of a function 193

13.1 Steps to plot a function 193

13.2 Solved exercises 194

    13.2.1 $f(x) = \frac{\log x + 1}{\log x - 1}$ 194

    13.2.2 $f(x) = xe^x$ 196

13.3 $f(x) = logx - \frac{1}{2}x^2$ 197

13.4 $f(x) = e^x \cdot \sqrt[3]{x^2}$ 199

13.5 $f(x) = \frac{\log^2 x}{x^3}$ 201

13.6 $f(x) = e^{\frac{1}{x^2-1}}$ 203

13.7 $f(x) = \frac{x}{\log x}$ 204

14 Integrals   207

14.1 Reinann's definition of integral 207

14.2 Properties of the integral 211

14.3 Theorems on integrals 212

14.4 Table of "immediate" integrals 218

    14.4.1 "Almost immediate" integrals 219

14.5 Integration techniques 220

    14.5.1 Applying the rules of the derivatives 221

    14.5.2 Integration by parts 222

14.6 Improper integrals 223

15 Functions of several variables   225

15.1 Domain 225

15.2 The graph of functions depending on several variables 226

    15.2.1 Plotting the domain of a function $\mathbb{R}^2 \longrightarrow \mathbb{R}$ 227

15.3 Partial derivatives 228

15.4 Extrema of vectorial functions 230

Bibliography   **238**

Suggested websites   **240**

Introduction

This book is intended to provide the basics of mathematical calculus and the knowledge of the quantitative methods used for economic and social analysis, for applications to finance and statistics.

The basic knowledge needed to read the book are algebra of polynomials, basic knowledge of set theory, equations and inequalities of first and second degree, radicals, logarithm properties, logarithmical and exponential equations, geometry on the cartesian plane, trigonometric functions and their properties, trigonometrical equations.

Since in the past teaching years, checking my students, I have observed some lacks of knowledge concerning logarithms and exponential functions, the first chapter is spent to present these topics in a very intuitive and simple way. To this end many numerical examples are provided.

The second chapter is the true starting point of the course; it deals with the mathematical logic, which can be considered a stepping stone for next themes and for any mathematical concept.

The third and the fourth chapters introduce the concepts of functions and relations; here some economic examples are presented.

The topology of the real axis (sixth chapter) contains the necessary tools to face limits theory. Limits definitions are presented using topology, without any metric concept; in my opinion this is enough for students in the economic fields. A big effort has been done to present the concepts in an intuitive way: numeric examples are suggested to the reader to fully understand the computation rules.

Afterwards derivatives and integrals are presented.

Last chapter contains a glance on functions depending on several variables; this kind of functions is widely used in economics and the book contains the tools to face, at least, applications in a bachelor course of economics. Because of the complexity of this topic proofs are not presented.

The book contains many solved examples and some exercises whose solution is left to the reader; for further practise the reader may refer to any book from the secondary school; to this end some web sites are suggested at the end of the book.

Chapter 1

Logarithms and exponentials

1.1 Main definitions and properties

1.1.1 Power properties

We assume the reader already knows the concepts underlying the powers; here we present the power properties. It is necessary for the reader to handle these properties so, if you are not confident with them, my advice is to use numerical examples to better understand them.

Basic power properties

Let a and b be stricly greater than 0; then

$$a^x \cdot a^y = a^{x+y} \tag{1.1}$$

$$\frac{a^x}{a^y} = a^{x-y} \tag{1.2}$$

$$(a^x)^y = a^{x \cdot y} \tag{1.3}$$

$$(a \cdot b)^x = a^x \cdot a^y \tag{1.4}$$

$$a^0 = 1 \tag{1.5}$$

The reader can easily check the first 4 properties (even using a calculator if needed), but last one is a bit tricky. It is not possible to prove that $a^0 = 1$, but it make sense, and I want to show you an example to show you the opportunity to introduce this rule.

So what I'm wondering is:

$$\text{why is it convenient to assume } 3^0 = 1?$$

Let's consider the fraction $\frac{9}{9}$; clearly it is equal to 1. If now we consider property (1.2) we can write:

$$\frac{9}{9} = \frac{3^2}{3^2} = 3^{2-2} = 3^0$$

so I hope you can realize why it is convenient to assume that $3^0 = 1$. And we can assume that $a^0 = 1$ even if a is negative.

Pay a lot of attention to the case

$$0^0$$

It is not defined; this means that it is not possible to compute 0^0.

From the previous properties we can derive some more properties.

Some more power properties

Let a be a non-zero number and n a positive integer number; then

$$a^{-x} = \frac{1}{a^x} \tag{1.6}$$

$$a^{\frac{1}{n}} = \sqrt[n]{a} \tag{1.7}$$

The first one comes from property (1.2); let's consider the fraction $\frac{2}{16}$; of course we can simplify and obtain $\frac{1}{8} = \frac{1}{2^3}$. Applying the property we can write:

$$\frac{2}{16} = \frac{2^1}{2^4} = 2^{1-4} = 2^{-3}$$

Now it should be clear why it make sense to write $2^{-3} = \frac{1}{2^3}$.

The second property we have just presented comes from properties (1.1) and (1.3). Think to the following expression:

$$\left(\sqrt[3]{8}\right)^3 = (2)^3 = \boxed{8}$$

Now I want to show that we can obtain the same result applying properties (1.7) and (1.1):

$$\left(\sqrt[3]{8}\right)^3 = \left(8^{\frac{1}{3}}\right)^3 = 8^{\frac{1}{3}} \cdot 8^{\frac{1}{3}} \cdot 8^{\frac{1}{3}} = 8^{\frac{1}{3}+\frac{1}{3}+\frac{1}{3}} = 8^1 = \boxed{8}$$

We get the same result applying property (1.3):

$$\left(\sqrt[3]{8}\right)^3 = \left(8^{\frac{1}{3}}\right)^3 = 8^{\frac{1}{3}\cdot 3} = 8^1 = \boxed{8}$$

Applying previous properties it should be clear to the reader that

$$\left(\frac{a}{b}\right)^x = \frac{a^x}{b^x}$$
$$\left(\frac{a}{b}\right)^{-1} = \frac{b}{a}$$
$$\left(\frac{a}{b}\right)^{-x} = \frac{b^x}{a^x}$$
$$a^{\frac{k}{n}} = \sqrt[n]{a^k}$$
$$a^{-\frac{k}{n}} = \frac{1}{\sqrt[n]{a^k}}$$

About last equalities I want to remark that it is not necessary to memorize (to store in some locations of our memory) these formulae; we can derive them just applying the previous properties.

1.1.2 Logarithm definition and properties

The expression $\log_a b$ reads logarithm base a of b (a is called *base* and b is called *argument* or *input*) and indicates the exponent we have to raise a to obtain b. This means that we can write the expression:

$$\log_a b = \boxed{?}$$

as

$$a^{\boxed{?}} = b$$

Example 1.1 $\log_2 8 = 3$ *since* $2^3 = 8$

$\log_5 25 = 2$ *since* $5^2 = 25$

$\log_6 \frac{1}{6} = -1$ *since* $6^{-1} = \frac{1}{6}$

$\log_8 2 = \frac{1}{3}$ *since* $8^{\frac{1}{3}} = \sqrt[3]{8} = 2$

Logarithms properties

Let a, b, c, x, y be strictly positive numbers, a, b, c different by 1 then

$$\log_a (xy) = \log_a x + \log_a y \tag{1.8}$$

$$\log_a \frac{x}{y} = \log_a x - \log_a y \tag{1.9}$$

$$\log_a x^k = k \cdot \log_a x \tag{1.10}$$

$$\log_a x = \frac{\log_c x}{\log_c a} \tag{1.11}$$

These properties are a direct consequences of the properties of the powers (the logarithms are nothing but exponents).

Example 1.2 *Consider* $\log_2 64$; *by direct computation we know that* $\log_2 64 = \boxed{6}$; *applying (1.8) we can write:*

$$\log_2(8 \cdot 8) = \log_2 8 + \log_2 8 = 3 + 3 = \boxed{6}$$

Considering now $\log_2 \frac{1}{4}$ *which is equal to* $\boxed{-2}$ *since* $2^{-2} = \frac{1}{2^2} = \frac{1}{4}$; *applying (1.9) we can write:*

$$\log_2 \frac{1}{4} = \log_2 1 - \log_2 4 = 0 - 2 = \boxed{-2}$$

Let's come back to $\log_2 64$ *and let's apply (1.10):*

$$\log_2 64 = \log_2 8^2 = 2\log_2 8 = 2 \cdot 3 = \boxed{6}$$

In my opinion property (1.11) is the less helpful one; it may be useful because some calculators and some computer softwares are able to compute only $\log_{10} x$ *and* $\log_e x$[1].

So if we want to compute $\log_2 3$ *we can write:*

$$\log_2 3 = \frac{\log_{10} 3}{\log_{10} 2} \approx 1.585$$

Remark 1.1 *Property (1.10) is the one that will be used more often in this book and in many economic applications, so don't forget it!*

From the properties of the powers, moreover, it is easy to deduce:

$$\log_a a = 1$$
$$\log_a 1 = 0$$
$$\log_a \frac{1}{a} = -1$$
$$\log_a b = \frac{1}{\log_b a}$$

---

[1] e is a number close to 3 ($e \approx 2.7$) which has many nice properties; some of theese properties will be discussed later in this book.

Once more I recommend you: don't make any effort to remember these properties, deeply understand them and make some numerical examples to be confident with them.

The following formulae linking exponential and logarithms are the most important ones in this chapter:

$$\boxed{\log_a a^x = x \quad \forall x}$$
$$\boxed{a^{\log_a x} = x \quad \forall x > 0}$$
(1.12)

We'll discuss these two relations in details later in this book.

1.2 Exponential Functions

The following picture represents the graph of the function $y = e^x$ (let me remind you that $e \approx 2.7$).

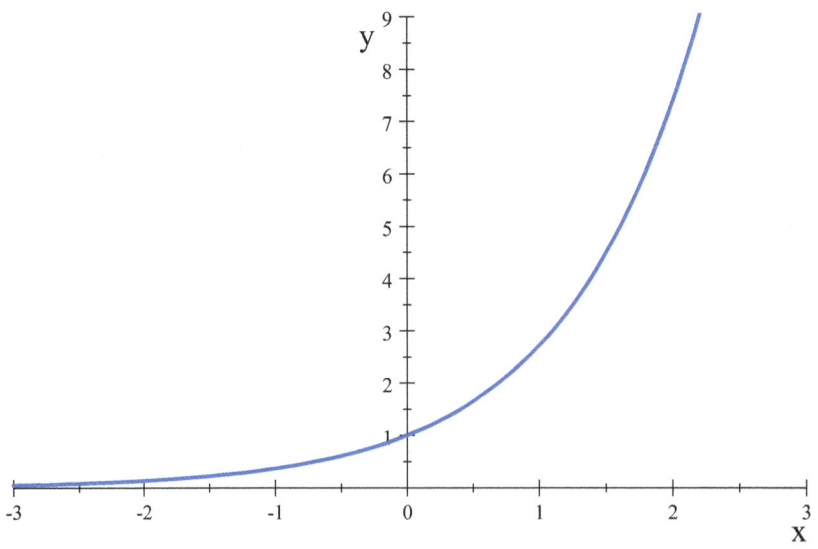

The shape depicted in the picture above is, somehow, the shape of any exponential function whose base a is greater than 1.

Exercise 1.1 *Represent on a single cartesian plane the functions $y = 2^x$ and $y = 3^x$. Before plotting complete the following table (use a calculator if needed):*

x	2^x	3^x
0		
1		
2		
3		
-1		
-2		
-3		

Any exponential function with base $a > 1$ satisfies the following properties:

1. The point $(0;1)$ always belongs to its graph;

2. $a^x > 0 \quad \forall x$ (any exponential function is always strictly positive);

3. $\lim\limits_{x \to +\infty} a^x = +\infty \qquad \lim\limits_{x \to -\infty} a^x = 0$

    We'll define the limits later in this book, here we are just introducing the concept of limit for such functions; writing $\lim\limits_{x \to +\infty} a^x = +\infty$ we mean that when the x becomes bigger and bigger ($x \longrightarrow +\infty$) the function becomes bigger and bigger; writing $\lim\limits_{x \to -\infty} a^x = 0$ we mean that when x becomes more and more negative ($x \longrightarrow -\infty$) the function becomes closer and closer to 0.

4. They are strictly increasing functions:

$$[x_1 < x_2] \iff [a^{x_1} < a^{x_2}]$$

In next picture we have the graph of an exponential function whose base a is positive but smaller than 1.

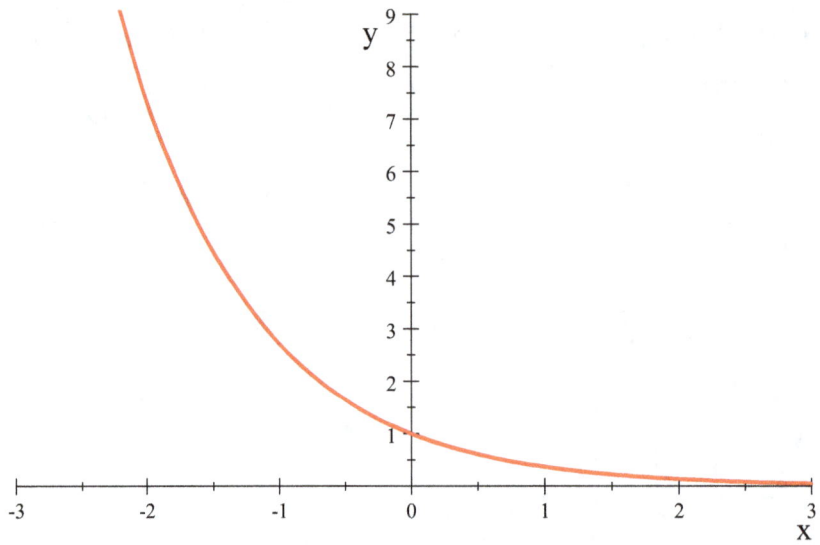

Exercise 1.2 *Represent on a single cartesian plane the functions $y = 0.5^x$ and $y = \left(\frac{1}{3}\right)^x$. Before plotting complete the following table (you can use a calculator):*

x	0.5^x	$\left(\frac{1}{3}\right)^x$
0		
1		
2		
3		
-1		
-2		
-3		

For this class of exponential functions (with $0 < a < 1$) it holds:

1. The point $(0; 1)$ always belongs to its graph;

2. $a^x > 0 \quad \forall x$ (an exponential function is never negative);

3. $\lim\limits_{x \longrightarrow +\infty} a^x = 0 \qquad \lim\limits_{x \longrightarrow -\infty} a^x = +\infty$

   Now writing $\lim\limits_{x \longrightarrow +\infty} a^x = 0$ we mean that when the x becomes bigger and bigger ($x \longrightarrow +\infty$) the function becomes closer and closer to 0; writing $\lim\limits_{x \longrightarrow -\infty} a^x = +\infty$ we mean that when x becomes more and more negative ($x \longrightarrow -\infty$) the function becomes bigger and bigger.

4. They are strictly decreasing functions:

$$[x_1 < x_2] \iff [a^{x_1} > a^{x_2}]$$

Example 1.3 *Solving logarithmic inequalities. Keep in mind (1.12) and property number 4 of exponential functions.*

$$\log_2 (x-1) > 3$$
$$\updownarrow$$
$$2^{\log_2(x-1)} > 2^3$$
$$\updownarrow$$
$$x - 1 > 8$$

$$\log_{0.1} (x-1) < 3$$
$$\updownarrow$$
$$0.1^{\log_{0.1}(x-1)} > 0.1^3$$
$$\updownarrow$$
$$x - 1 > 0.001$$

1.3 Logarithmic Functions

In the previous section we said that exponential functions are always positive. This implies that $3^x > 0$ for every x.

For this reason we cannot compute the logarithm with base 3 of a negative number; hence the (fundamental) condition for the existence of a logarithm is: the argument (input) must be greater than zero.

Moreover since $1^x = 1 \forall x$ it is not possible to compute $log_1 x$: let's imagine we try to compute $\log_1 3$, we'll never find (it is not possible to determine) an exponent w such that $1^w = 3$.

Summarizing: $\log_a x$ does exist when $x > 0$ and $a \neq 1$.

Next picture shows the shape of function $y = \log_a x$ when $a > 1$.

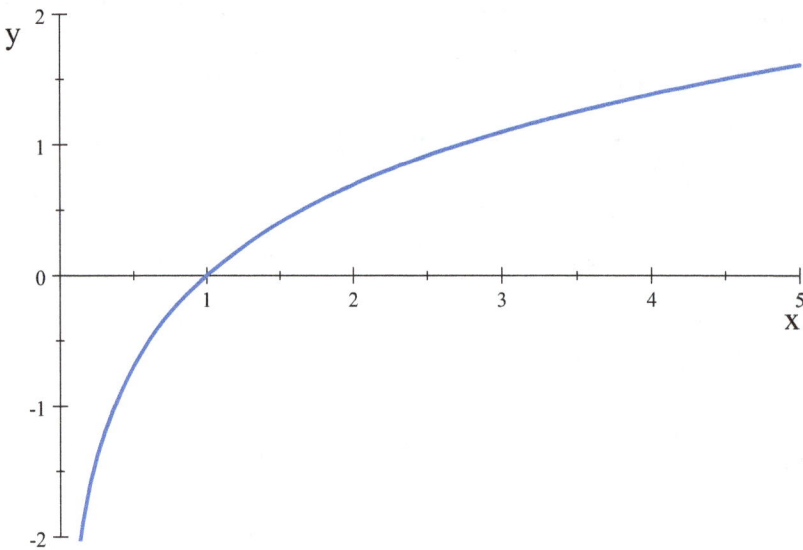

Picture shows the shape of any logarithmic function with base greater than 1.

Exercise 1.3 *Represent on a single cartesian plane the functions $y = \log_2 x$ and*

$y = \log_4 x$. Before plotting complete the following table (use a calculator if needed):

x	$\log_2 x$	$\log_4 x$
1		
2		
4		
16		
1/2		
1/4		
1/16		

For these functions the following properties hold:

1. The point $(1; 0)$ belongs to their graph;

2. $log_a x$ exists if and only if $x > 0$;

3. $log_a x > 0$ when $x > 1$;

4. $\lim_{x \to +\infty} \log_a x = +\infty \qquad \lim_{x \to 0^+} \log_a x = -\infty$

    The first limit means that when x becomes bigger and bigger ($x \to +\infty$) the logarithm becomes bigger and bigger. The second one means that when the x gets closer and closer to 0 from the right ($x \to 0^+$) the function becomes more and more negative.

5. They are strictly increasing functions:

    $$[0 < x_1 < x_2] \iff [\log_a x_1 < \log_a x_2]$$

In next picture we consider a logarithm with base between 0 and 1.

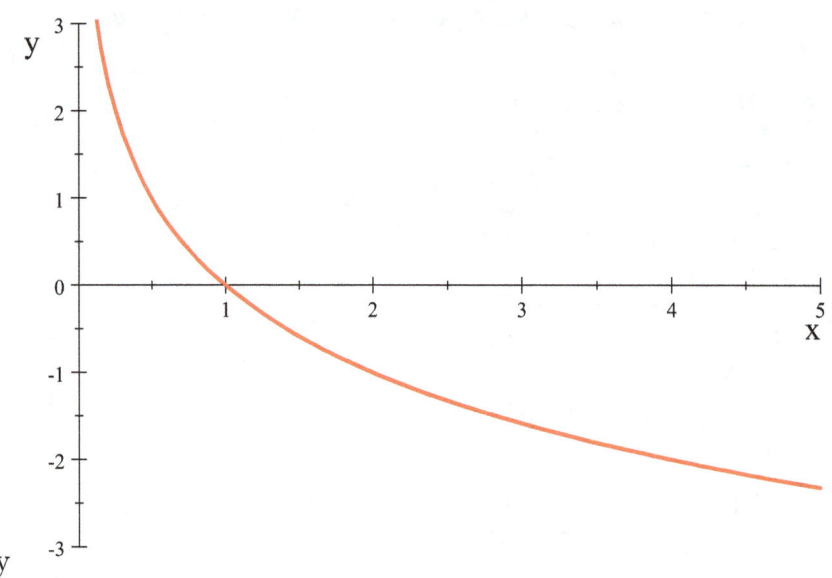

Every function $y = \log_a x$ when $0 < a < 1$ has a shape similar to the one above.

Exercise 1.4 *Represent on a single cartesian plane the functions $y = \log_{0.5} x$ and $y = \log_{\frac{1}{4}} x$. Before plotting complete the following table (use a calculator if needed):*

x	$\log_{0.5} x$	$\log_{\frac{1}{4}} x$
1		
2		
4		
1/2		
1/4		
1/16		

They satisfy the following properties:

1. The point $(1; 0)$ belongs to their graph;

2. $log_a x$ exists if and only if $x > 0$;

3. $\log_a x > 0$ when $0 < x < 1$;

4. $\lim\limits_{x \to +\infty} \log_a x = -\infty \qquad \lim\limits_{x \to 0^+} \log_a x = +\infty$

   The first limit means that when x becomes bigger and bigger ($x \to +\infty$) the logarithm becomes more and more negative. The second one means that when the x gets closer and closer to 0 from the right ($x \to 0^+$) the function becomes bigger and bigger.

5. They are strictly decreasing functions:

$$[0 < x_1 < x_2] \iff [\log_a x_1 > \log_a x_2]$$

Example 1.4 *Solving exponential inequalities. Keep in mind (1.12) and property number 5 of logarithmic functions.*

$$3^{x-1} > 9$$
$$\updownarrow$$
$$\log_3 3^{x-1} > \log_3 9$$
$$\updownarrow$$
$$x - 1 > 2$$

$$\left(\frac{1}{2}\right)^x < 4$$
$$\log_{\frac{1}{2}} \left(\frac{1}{2}\right)^x > \log_{\frac{1}{2}} 4$$
$$x > -2$$

I remind you that when using calculators $\boxed{ln}$ is the logarithm with base e (it is also called natural logarithm); $\boxed{\log}$ instead is the logarithm with base 10. This notation is not standard: some books use log for natural logarithms and *Log* for logarithms with base equal to 10.

In most books dealing with financial applications log denotes the natural logarithm; for this reason I'm following this notation. In any case in this course we'll never use logarithms with base equal to 10.

Chapter 2

Boolean algebra

2.1 Statements and operators

The basic elements of the boolean algebra are statements (or sentences or propositions). Statements in everyday language are somehow ambigous in the sense that sometimes it is not possible to determine if they are true or false, or they may be true for somebody but false for somebody else (think as an example to the statement "Giovanni Quaranta has glamour"). In this algebra statements are not ambigous and they can be only true (T or 1) or false (F or 0), even if we don't know if they are true or false.

We have two kinds of statements:

unconditional statements: they are propositions assuming always the same value (ex. "3 is an even number"); we denote unconditional statements using capital letters in italics (P).

statements depending on one (or more) **variable**(s): their value depends on an argument (ex. "x is an even number"); in this case we can say if the statement is true only if we know x. We denote statements depending on variables using capital letters in italics and writing the argument inside brackets ($P(x)$).

We can build complex statements using logic operators (also called connectors); the basic ones are:

negation: **not** ($\neg$)

disjunction: **or** ($\vee$)

conjunction: **and** ($\wedge$)

implication: $\implies$

equivalence $\iff$.

2.2 Truth Tables

In this section we analyze the rules underlying the fundamental boolean operators.

Let P and Q be two statements then:

$not(P)$ (or simply $notP$) is false when P is true, and is true if P is false. We can summaryze the behavior of the operators using truth tables; in the case of the not operator we have:

P	$not(P)$
1	0
0	1

Let's consider: $P :=$ "3 is an even number". Of course the sentence is false but the statement $not(P)$ which means "3 is not an even number" is clearly true.

(P or Q) is true if at least one between P and Q is true; false when both are false.

Let:

$P(x) :=$ "x is an even number",

$Q(x) :=$ "x is a multiple of 5".

The statement $(P(x) \text{ or } Q(x))$ can be expressed as "x is multiple of 2, x is a multiple of 5, or x is a multiple of 2 and 5 at the same time".

Exercise 2.1 *Considering the proposition $P(x)$ and $Q(x)$ as above, determine if the statement $(P(x) \text{ or } Q(x))$ is true or not when x is one of the following numbers: 20, 7, 15, 8.*

The corresponding truth table is

P	Q	P or Q
1	1	**1**
1	0	**1**
0	1	**1**
0	0	**0**

It is important to notice that in this case the table has 4 rows (excluding the heading). This happens because when building a truth table we have to consider all the possible cases; now we have 2 sentences and each one can assume two possible values hence the number of all possible cases is $2^2 = 4$.

It is also important to remark that when doing an exercise it is helpful to use always the same order for the values in the first two columns.

$(P \text{ and } Q)$ is true when both statements are true, and false in any other case.

Let:

$P(x) :=$ "x is an even number",

$Q(x) :=$ "x is a multiple of 5".

The statement $(P(x)$ and $Q(x))$ can be expressed as "x is (at the same time) multiple of 2 and multiple of 5" which means that "x is a multiple of 10".

Exercise 2.2 *Considering the proposition $P(x)$ and $Q(x)$ as above, determine if the statement $(P(x)$ and $Q(x))$ is true or not when x assumes one of the following values: 20, 7, 15, 8.*

The corresponding truth table is

P	Q	P and Q
1	1	1
1	0	0
0	1	0
0	0	0

$(P \implies Q)$ is true when P is false (whatever is Q) or when Q is true (whatever is P); implication is false only when P is true and Q is false.

Somehow $(P \implies Q)$ expresses the sentence "if P then Q". When writing $(P \implies Q)$ we say that:

P is the assumption,

Q is the consequence.

Moreover we say that:

P is a sufficient condition for Q,

Q is a necessary condition for P.

Example 2.1 *let $I :=$ "I am in Italy" and $E :=$ "I am in Europe". It is easy to detect the fairness of the sentence:*

$$I \implies E$$

since if I am in Italy everybody can conclude that I am in Europe (so to be in Italy is a sufficient condition to be in Europe). The example also underlines that to be in Europe is a necessary (not sufficient) condition to be in Italy since if I am not in Europe I cannot be in Italy; moreover if I am in Europe I may be in Italy or not.

The truth table is:

P	Q	$P \implies Q$
1	1	**1**
1	0	**0**
0	1	**1**
0	0	**1**

Remark 2.1 *It is important to realize that when the assumption is false the implication is always true (consider last two rows of the table). Don't forget!*

$(P \iff Q)$ is true when the propositions have the same value, i.e. when both are true or both are false; it is false when they have different values. In this case we say that P is a necessary and sufficient condition for Q.

We have the following truth table:

P	Q	$P \iff Q$
1	1	**1**
1	0	**0**
0	1	**0**
0	0	**1**

When considering a complex statement keep in mind that the not operator is the one which has to be considered before any other operator; in this text I will use brackets to avoid ambiguity about the order to be followed.

2.2.1 Proving the fairness of a statement

We can use truth tables to prove the fairness of complex statements; let's consider the following example.

Example 2.2 *Let's prove that*

$$[P \implies Q] \iff [not\,(P)\ or\ Q]$$

to achieve our goal we can use a truth table listing the set of all possible cases and showing that the equivalence is true in any case.

P	Q	$[P \implies Q]$ $\boxed{1}$	$\iff$ $\boxed{4}$	$not\,(P)$ $\boxed{2}$	or $\boxed{3}$	$Q]$
1	1	1	1	0	1	1
1	0	0	1	0	0	0
0	1	1	1	1	1	1
0	0	1	1	1	1	0

The number in a box under each operator is the order followed while building the table: $[P \implies Q]$ is the first one, $not\,(P)$ the second one and so on. The column in red is last one that has been computed; it shows the equivalence between the two columns in blue.

Since the equivalence is always true (we have 1 everywhere in the corresponding column), the proposition presented is correct.

2.3 Tautologies and contradictions

Definition 2.1 *A **tautology** is a statement always true.*

*A **contradiction** is a statement always false.*

The complex statement in example 2.2 is a tautology.

A tautology can be considered as a rule, as something that is always correct.

Here are some easy tautologies:

$$P \implies P$$

$$P \iff P$$

$$P \text{ or } not\,(P)$$

$$P \iff not\,(not\,(P))$$

A simple example of contradiction is:

$$P \text{ and } not\,(P)$$

My advice is to check the fairness of some of theese tautologies but it is not necessary to remember them.

Two important tautologies are the wellknown De Morgan's laws:

$$[not\,(P \text{ and } Q)] \iff [notP \text{ or } notQ] \qquad (2.1)$$

$$[not\,(P \text{ or } Q)] \iff [notP \text{ and } notQ] \qquad (2.2)$$

In order to better present the equivalence (2.1) let's consider:

$P(x) :=$ "x is an even number",

$Q(x) :=$ "x is a multiple of 5".

As already seen the sentence $R(x) = P(x)$ and $Q(x)$ means "x is a multiple of 10";

$not\,(P\ and\ Q)$ is the same as to say $not\,R(x)$, that is "x is not a multiple of 10" and this happens if x is not even ($not\,P(x)$) **or** x is not a multiple of 5 ($not\,Q(x)$).

A very important (very very important) tautology is the contradiction principle that will be widely used later in this book:

$$\boxed{[P \Longrightarrow Q] \Longleftrightarrow [not\,P \Longrightarrow not\,Q]}$$

Exercise 2.3 *Using the truth tables prove the De Morgan's laws and the contradiction principle.*

Boolean operators *and* and *or* satisfy other laws which are quite similar to the rule of the standard algebra for the operators $+$ and $\cdot$:

Commutative properties:

$$[P\ and\ Q] \Longleftrightarrow [Q\ and\ P]$$
$$[P\ or\ Q] \Longleftrightarrow [Q\ or\ P]$$

Associative properties:

$$[(P\ and\ Q)\ and\ R] \Longleftrightarrow [P\ and\ (Q\ and\ R)]$$
$$[(P\ or\ Q)\ or\ R] \Longleftrightarrow [P\ or\ (Q\ or\ R)]$$

Distributive properties:

$$[P \text{ and } (Q \text{ or } R)] \iff [(P \text{ and } Q) \text{ or } (P \text{ and } R)] \tag{2.3}$$

$$[P \text{ or } (Q \text{ and } R)] \iff [(P \text{ or } Q) \text{ and } (P \text{ or } R)] \tag{2.4}$$

Remark 2.2 *In boolean algebra it is correct to distribute the and operator with respect to the or as in (2.3) and the or operator with respect to the and; in the standard algebra it is possible to distribute the · operator with respect to the + and hence it is correct to write:*

$$a \cdot (b + c) = (a \cdot b) + (a \cdot c)$$

while it is not possible to distribute the + with respect to ·; hence in general

$$a + (b \cdot c) \neq (a + b) \cdot (a + c)$$

(Use numerical examples if you are not confident with this property).

It is not necessary to remember all these equivalences; it may be useful to verify some of them using the truth tables. Keep in mind that when three propositions are involved you need $2^3 = 8$ rows (excluding the headings) as in the following example (the numbers in the boxes are to suggest the order to follow):

P	Q	R	[P	or	(Q and R)]	$\iff$	[(P or Q)	and	(P or R)]
				2	1	6	3	5	4
1	1	1							
1	1	0							
1	0	1							
1	0	0							
0	1	1							
0	1	0							
0	0	1							
0	0	0							

Exercise 2.4 *I define a new operator aut (Latin) or xor that somehow meets the "either...or" in the everyday language; the "or" operator is true when both statements are true, while, when speaking, we often use the "or" in an exclusive sense:*

"either I go to the cinema or I stay home" meaning that I cannot do both.

Here is its truth table:

P	Q	P xor Q
1	1	**0**
1	0	**1**
0	1	**1**
0	0	**0**

Find one (or more than one) proposition corresponding to P xor Q.

Solution 2.1 *the following are some possible solutions to the exercise*

$\text{not}\,(P \iff Q)$

$(P \text{ or } Q) \text{ and not}\,(P \text{ and } Q)$

$(P \text{ and not}Q) \text{ or } (\text{not}P \text{ and } Q)$

$\text{not}\,(P \implies Q) \text{ or not}\,(Q \implies P)$

2.3.1 What is really necessary?

Every compounded sentence can be obtained using only the *not* and the *or* operators. The other operators are somehow redundant.

We already saw in example 2.2 that the connector $\implies$ can be expressed using *not* and *or* operators; in fact using truth tables it is easy to prove that

$(P \text{ and } Q) \iff \text{not}\,[\text{not}P \text{ or not}Q]$

$(P \iff Q) \iff [(P \implies Q) \text{ and } (Q \implies P)]$

2.4 Quantifiers

Universal quantifier:

∀ "for all"

Existential quantifier:

∃ "there exists at least one"

Moreover sometimes will use another quantifier:

∃! "there exists only one".

Let's think to the negation of the following statement:

$$\text{ALL BOTTLES ARE FULL}$$

Then not(ALL BOTTLES ARE FULL) is equivalent to:

$$\text{ALL BOTTLES ARE EMPTY}$$

or

$$\text{THERE IS AT LEAST ONE BOTTLE THAT IS NOT FULL?}$$

The right form of the negation is the second one!

I use this example to show that the negation of the sentence:

$$\forall x, x \text{ satisfies a property}$$

is:

$\exists x$ that doesn't satisfy that property

Example 2.3 *Let $A := \{1, 2, 3, 5, 7, 9\}$ then:*

$\forall x \in A$, x is odd (all the elements of A are odd) is false;

$\exists x \in A : x$ is not odd (at least one element in A is not odd) is true.

Chapter 3

Functions

3.1 Basic definitions

Definition 3.1 *The cartesian product between two sets A and B is the set of all the possible pairs obtained taking the first element in A and the second one in B:*

$$A \times B \stackrel{def}{=} \{(a;b) : a \in A, b \in B\}$$

Example 3.1 *Let $A := \{1,2,3\}$ and $B := \{a,b\}$ then*

$$A \times B = \{(1;a),(1;b),(2;a),(2;b),(3;a),(3;b)\}$$

Note that $(a;1)$ does not belong to $A \times B$ (it belongs to $B \times A$).

Definition 3.2 *A binary relation $\mathcal{R}$ from A to B ($A \longrightarrow B$) is a subset of $A \times B$.*

To specify that $(a;b)$ belongs to a relation we write: $(a;b) \in \mathcal{R}$ or $a\mathcal{R}b$; we'll use the notation $a\not\mathcal{R}b$ to denote that $(a;b) \notin \mathcal{R}$.

Definition 3.3 *A function from A to B is a relation which matchs every element in A with a single element in B.*

We can say that a relation from A to B is a rule that assigns to some elements of A some elements of B (one, no one or infinity many ones); a function is a rule that at **each** elements of A assigns **only one** element of B.

Let $f : A \longrightarrow B$ be a function; we call the set A domain and the set B range (sometimes also called codomain). We'll often refer to the domain of f writing D_f.

The notation used to specify that b is the element assigned to a by the function f is:

$$b = f(a) \quad \text{or} \quad f : a \longmapsto b$$

and we say that b is the image of a; we call a inverse image of b and we write:

$$a = f^{-1}(b)$$

The notation $y = f(x)$ is widely used to denote that variable y depends on variable x; x is said independent variable and y is the dependent one.

In the first part of the course we'll deal with functions depending on a single variable (like $y = \sin x$). In the last part we'll face functions depending on two or more variables (ex. $z = \sin(x+y)$).

The graph of a function $f : A \subseteq \mathbb{R} \longrightarrow B \subseteq \mathbb{R}$ is the subset of $\mathbb{R}^2$ of the points belonging to the relation (a function is a relation); we will use the notation Γ_f to denote the grapf of f:

$$\Gamma_f \stackrel{def}{=} \{(x; y) \in \mathbb{R}^2 : y = f(x)\}$$

In the cartesian plane the domain is always a subset of the horizontal axis, the domain a subset of the vertical one.

We call upper graph of f the set: $\Gamma_f^+ \stackrel{def}{=} \{(x;y) \in \mathbb{R}^2 : y \geq f(x)\}$

We call lower graph of f the set: $\Gamma_f^- \stackrel{def}{=} \{(x;y) \in \mathbb{R}^2 : y \leq f(x)\}$

Example 3.2 *Let's consider the function $y = x^2 + 1$ $\mathbb{R} \longrightarrow \mathbb{R}$; the following picture represents its graph (the blue curve), its upper graph (the yellow area) and the lower graph (the grey area):*

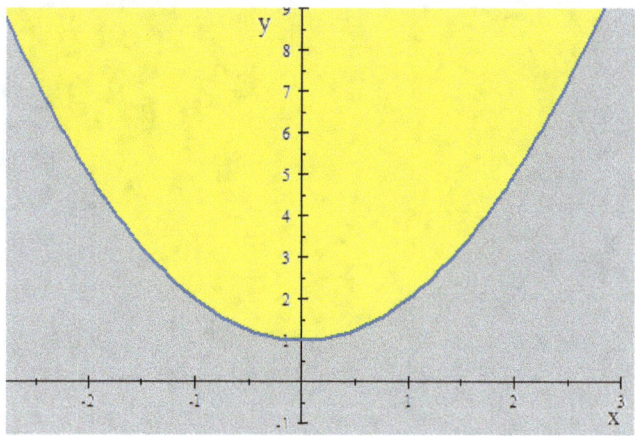

3.2 Further definitions

Given any function $f : A \longrightarrow B$, we can define the image of a subset X of A:

$$f(X) \stackrel{def}{=} \{y \in B : y = f(x) \, \forall x \in X\}$$

and the inverse image of a subset Y of B:

$$f^{-1}(Y) \stackrel{def}{=} \{x \in A : y = f(x) \, \forall y \in Y\}$$

Definition 3.4 *A function $f : A \longrightarrow B$ is said surjective if $f(A) = B$ that is the image of the domain is the whole range.*

This means that every element in the range is image of some elements in the domain:

$$\forall y \in B \, \exists x \in A : y = f(x)$$

Example 3.3 *The picture shows a representation of a surjective function; domain is the set on the right hand side, range the one on the left:*

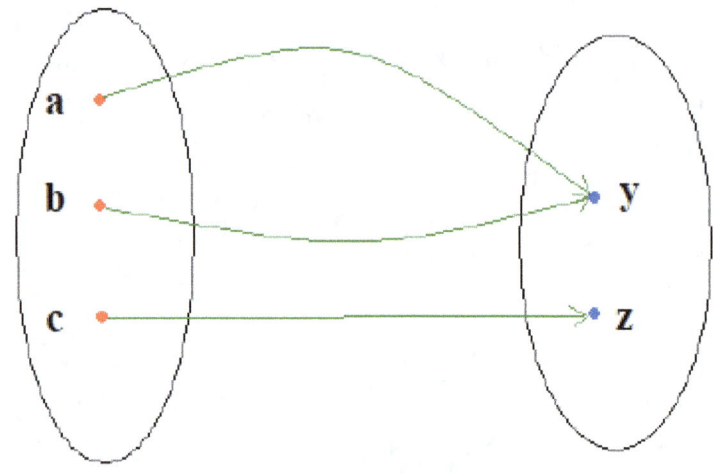

in this case we have:

$$f(a) = y \qquad f(b) = y \qquad f(c) = z;$$

$$f^{-1}(y) = \{a, b\} \qquad f^{-1}(z) = c.$$

Here is a representation of a non surjective function:

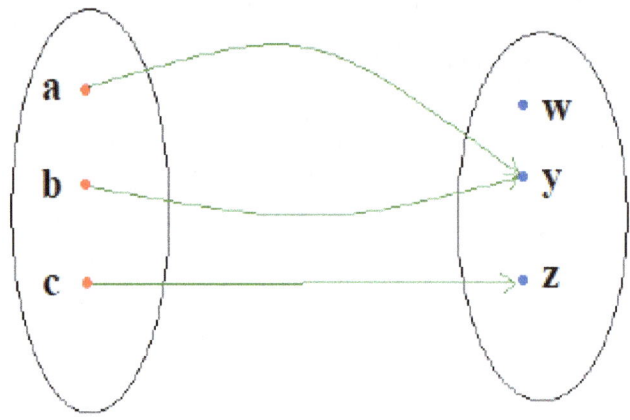

In this case the inverse image of w is the empty set: $f^{-1}(w) = \emptyset$

Definition 3.5 *A function is injective if different elements of the domain have different images; we can write:*

$$(a_1 \neq a_2) \implies [f(a_1) \neq f(a2)]$$

or (calling down the contradiction principle)

$$[f(a_1) = f(a2)] \implies (a_1 = a_2)$$

Example 3.4 *Here is a graphical example of an injective function (not surjective)*

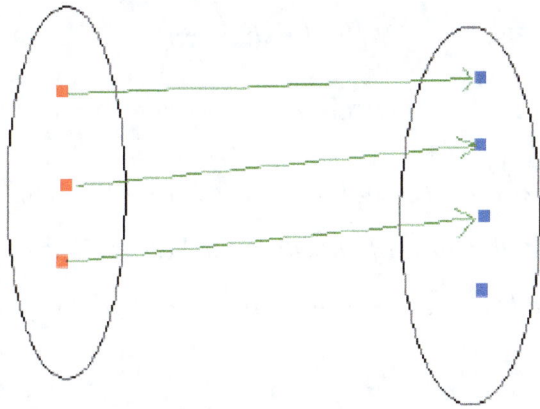

Here is an example of a non injective function (it is surjective)

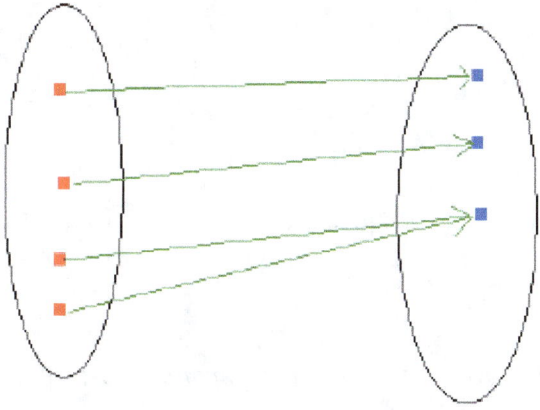

Definition 3.6 *A function is a one to one correspondence if it is injective and surjective.*

Proposition 3.1 *It is possible:*

1. *to make surjective a function cutting the range;*

2. *to make injective a function cutting the domain.*

Example 3.5 *vertical straight lines are not considered functions.*

Horizontal lines (like $y = 3$) are not surjective if the range is $\mathbb{R}$.

Straight lines that are neither horizontal nor vertical are one to one correspondences.

Exponential functions $\mathbb{R} \longrightarrow \mathbb{R}$ are injective but not surjective; from the graphs below it is possible to check that different elements of the domain have different images (injectivity) but 0 and negative numbers in the range are image of nothing (non surjectivity).

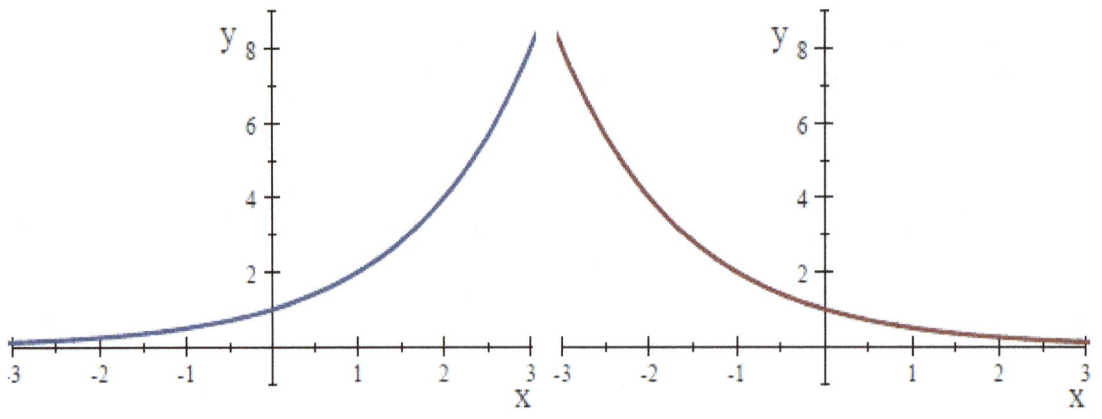

Exponential functions $\mathbb{R} \longrightarrow \mathbb{R}_+ - \{0\}$ are one to one correspondences; note that the range has been cut in order to make the function surjective.

Logarithmic functions $\mathbb{R}_+ - \{0\} \longrightarrow \mathbb{R}$ are one to one correspondences (see pictures below).

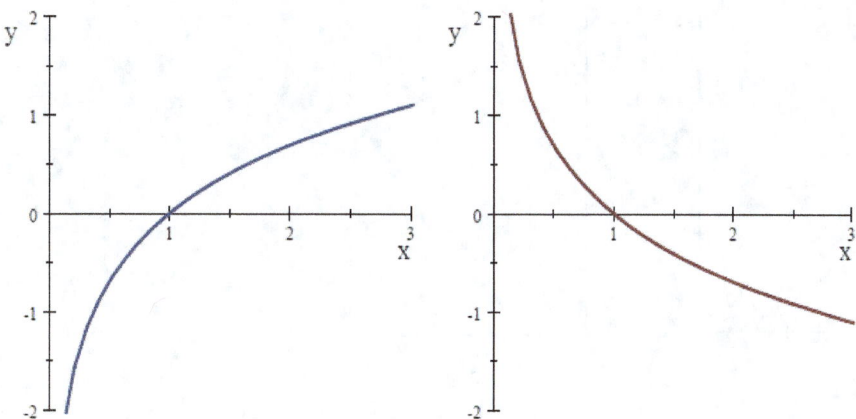

Power functions with even exponent $\mathbb{R} \longrightarrow \mathbb{R}$ are neither injective nor surjective; from the picture below you can detect non-injectivity: considering, as an example, element 4 in the range, it is image of two different elements in the domain (± 2) so the function is not injective; considering negative numbers in the range we can see that they are image of nothing.

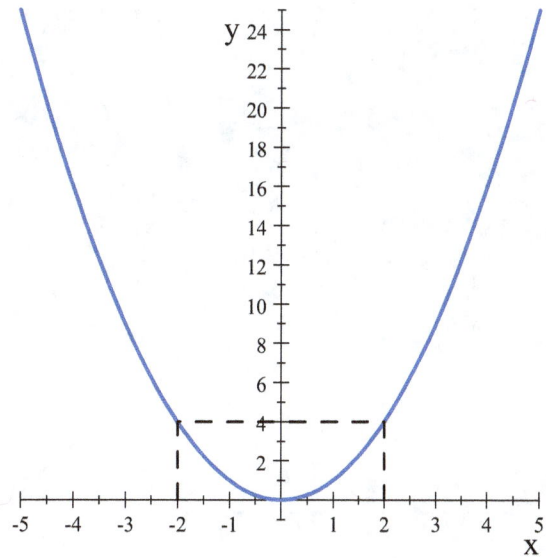

Power functions with even exponent $\mathbb{R}_- \longrightarrow \mathbb{R}_+$ or $\mathbb{R}_+ \longrightarrow \mathbb{R}_+$ are one to one correspondences (consider graphs below and think over how domain and range have been cut).

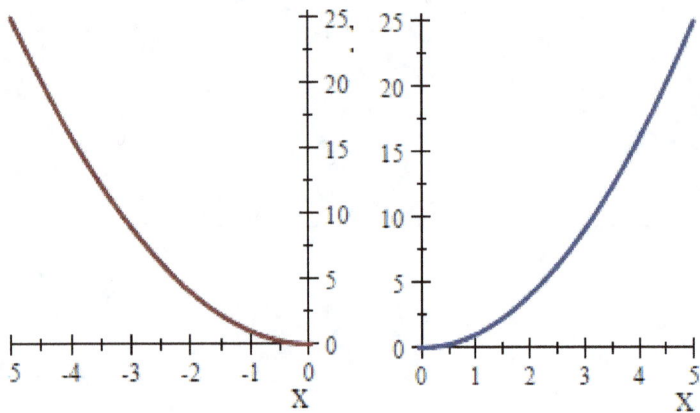

roots with even index $\mathbb{R}_+ \longrightarrow \mathbb{R}$ are injective but not surjective

roots with even index $\mathbb{R}_+ \longrightarrow \mathbb{R}_+$ are one to one correspondences

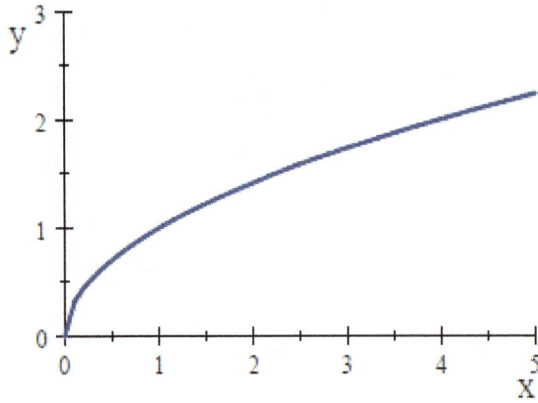

Power functions with odd exponent $\mathbb{R} \longrightarrow \mathbb{R}$ are one to one correspondences.

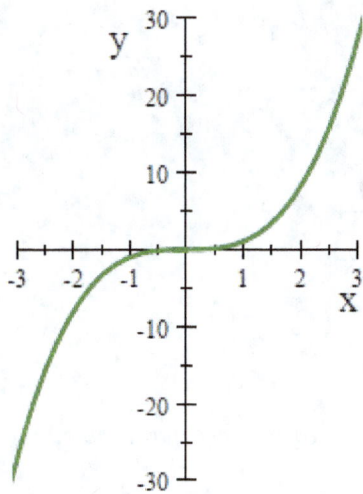

roots with odd index $\mathbb{R} \longrightarrow \mathbb{R}$ are one to one correspondences.

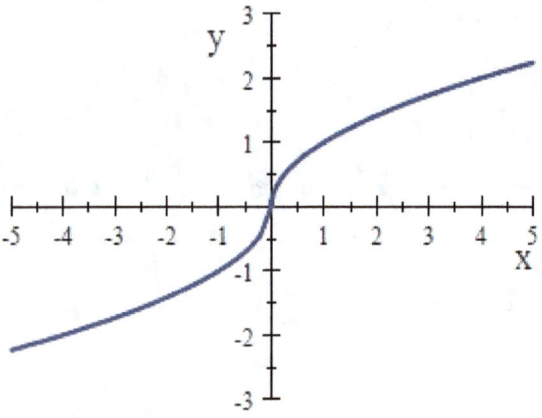

the function $\sin x : \mathbb{R} \longrightarrow \mathbb{R}$ is neither injective nor surjective. Think as an example that element 1 in the range is image of infinity many elements of the domain and so it is not injective; moreover elements in the range outside the interval $[-1; 1]$ are image of nothing and so the function is not surjective.

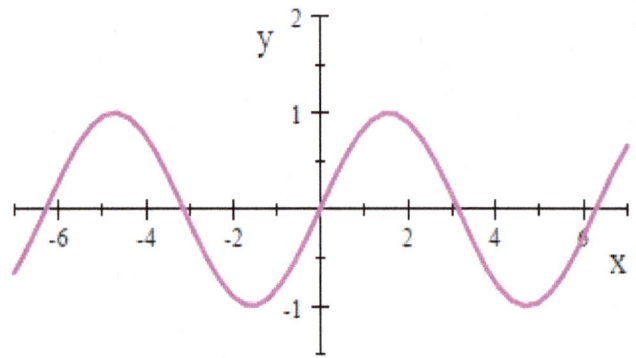

the function $\sin x : \left[-\frac{\pi}{2}; \frac{\pi}{2}\right] \longrightarrow [-1; 1]$ is a one to one correspondence

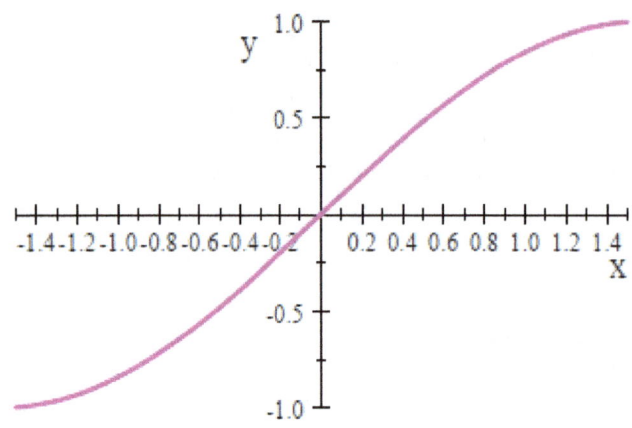

the function $\tan x : \mathbb{R} - \left\{(2k+1)\frac{\pi}{2}\right\}_{k \in \mathbb{N}} \longrightarrow \mathbb{R}$ is surjective but not injective.

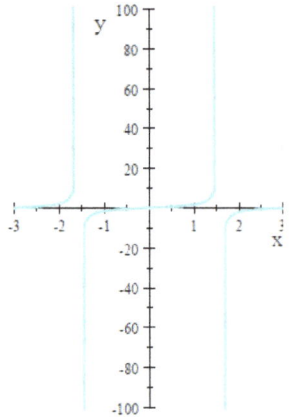

the function $\tan x : \left]-\frac{\pi}{2}; \frac{\pi}{2}\right[\longrightarrow \mathbb{R}$ is a one to one correspondence.

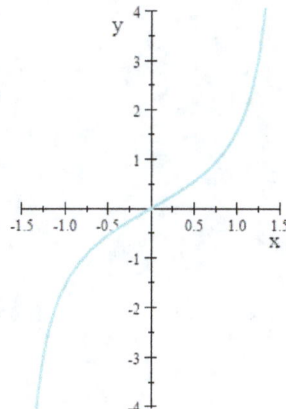

Exercise 3.1 *consider the function* $\cos x : \mathbb{R} \longrightarrow \mathbb{R}$ *and suggest how to cut domain and range in order to make it a one to one corresponce..*

3.3 Compound function (function of another function)

Let $g : A \longrightarrow B$ and $f : B \longrightarrow C$, we define the compound function of f and g:

$$(f \circ g)(x) = f(g(x)) : A \longrightarrow C$$

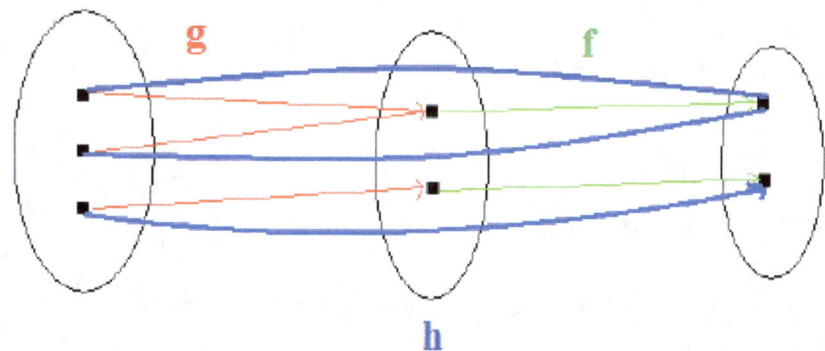

Example 3.6 *Let us consider the following functions:*

$f(x) = \log_2 x$ and $g(x) = \sin x$ then:

$(f \circ f)(x) = f(f(x)) = f(\log_2 x) = \log_2(\log_2 x)$

Let x be equal to 16 then:

$$\log_2(\log_2 16) = \log_2 4 = 2$$

$(g \circ g)(x) = g(g(x)) = g(\sin x) = \sin(\sin x)$

Let x be equal to π then:

$$\sin(\sin \pi) = \sin 0 = 0$$

$(g \circ f)(x) = g(f(x)) = g(\log_2) = \sin(\log_2 x)$

Let x be equal to 1 then:

$$\sin(\log_2 1) = \sin 0 = 0$$

$(f \circ g)(x) = f(g(x)) = f(\sin x) = \log_2(\sin x)$

Let x be equal to $\frac{\pi}{2}$ then:

$$\log_2(\sin x) = \log_2 1 = 0$$

Exercise 3.2 *Let's consider:*

$$f(x) = \sqrt{x^2 + 1}, \qquad g(x) = \frac{x+1}{x-1}, \qquad h(x) = x - a, \qquad k(x) = x + a$$

write the expressions of the compound functions:

$(f \circ f \circ ... \circ f)(x)$ *(f of f compounded n times)*

$(g \circ g)(x)$

$(h \circ h \circ ... \circ h)(x)$ *(h of h compounded n times)*

$(k \circ k \circ ... \circ k)(x)$ (k of k compounded n times)

$(h \circ k)(x)$

$(k \circ h)(x)$

3.4 Inverse functions

Definition 3.7 *Let $f : A \Longrightarrow B$ be a function; the inverse function noted here as g (or f^{-1}) the function $B \Longrightarrow A$:*

$$f(a) = b \Longrightarrow g(b) = a$$

the function $y = \sqrt[3]{x}$ is the inverse function of $y = x^3$: in the following picture you can see their graphs in the same cartesian plane:

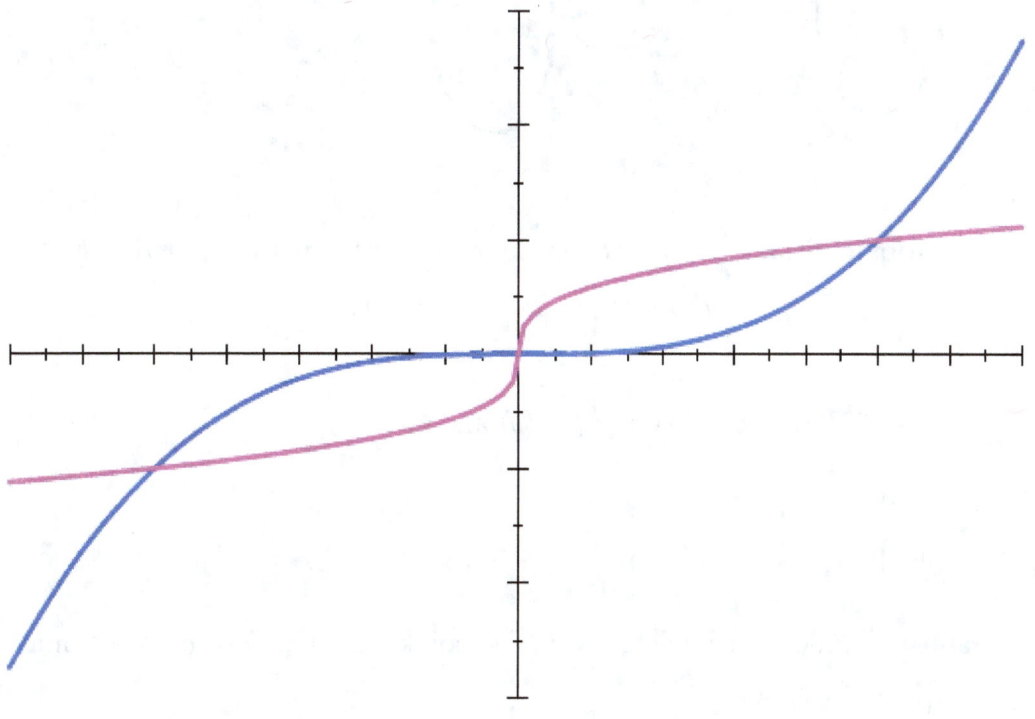

Remark 3.1 Let $f : A \longrightarrow B$ be a function; its inverse function g does exist if and only if f is a one to one correspondence. Indeed if f is not surjective the inverse relation wouldn't be a function since some points in the domain of the inverse function won't have an image; see the following representation:

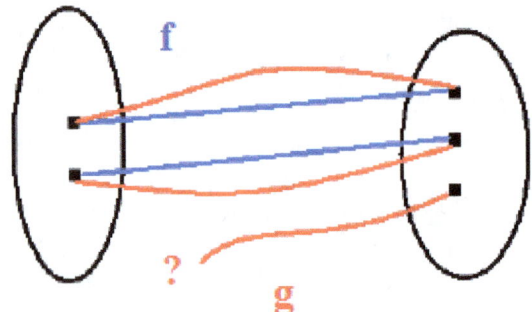

if f is not injective the inverse relation doesn't assign to an element a single output (so it is not a function):

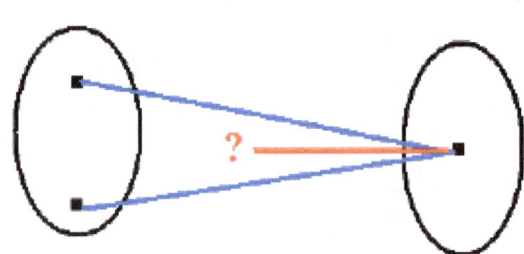

Proposition 3.2 we can obtain the graph of the inverse function by the graph of f with a symmetry with respect to the line $y = x$.

Proof. if $b = f(a)$ then $a = f^{-1}(b)$ and

$$\begin{cases} (a,b) \in \Gamma_f \\ (b,a) \in \Gamma_{f^{-1}} \end{cases}$$

and the proof is complete since the points (a, b) and (b, a) are symmetrical with respect to the line $y = x$. ∎

3.5 Inverse of the elementary functions

3.5.1 Power functions with odd exponent

Power functions with odd exponent $\mathbb{R} \longrightarrow \mathbb{R}$ are one to one correspondences and hence their inverse functions exist; their inverse functions are roots with (the same) odd index and they are functions $\mathbb{R} \longrightarrow \mathbb{R}$. Consider the picture below:

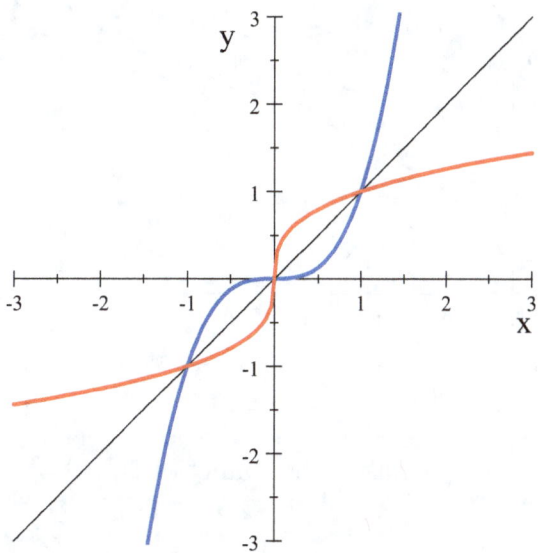

In this picture we have the graphs of $y = x^3$ (in blue) and $y = \sqrt[3]{x}$ (in red); the thin black line is the line whose equation is $y = x$, and I recommend you to take care of the symmetry.

3.5.2 Power functions with even exponent

These functions are not one to one correspondences (when we consider them $\mathbb{R} \longrightarrow \mathbb{R}$ they not are neither injective nor surjective); using suitable restrictions we can make them one to one correspondences and we have:

$f(x) = x^{2n} : \mathbb{R}_+ \longrightarrow \mathbb{R}_+$ has inverse function $f^1(x) = \sqrt[2n]{x} : \mathbb{R}_+ \longrightarrow \mathbb{R}_+$.

In the following picture we can view this case representing x^2 (in blue) and $\sqrt{x}$ (in red); once again it is easy to note the symmetry

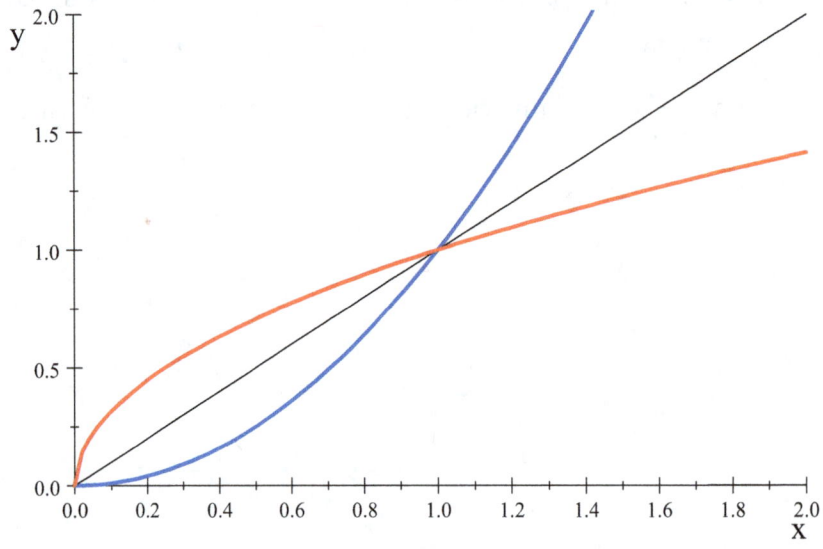

$f(x) = x^{2n} : \mathbb{R}_- \longrightarrow \mathbb{R}_+$ has inverse function $f^1(x) = -\sqrt[2n]{x} : \mathbb{R}_+ \longrightarrow \mathbb{R}_-$

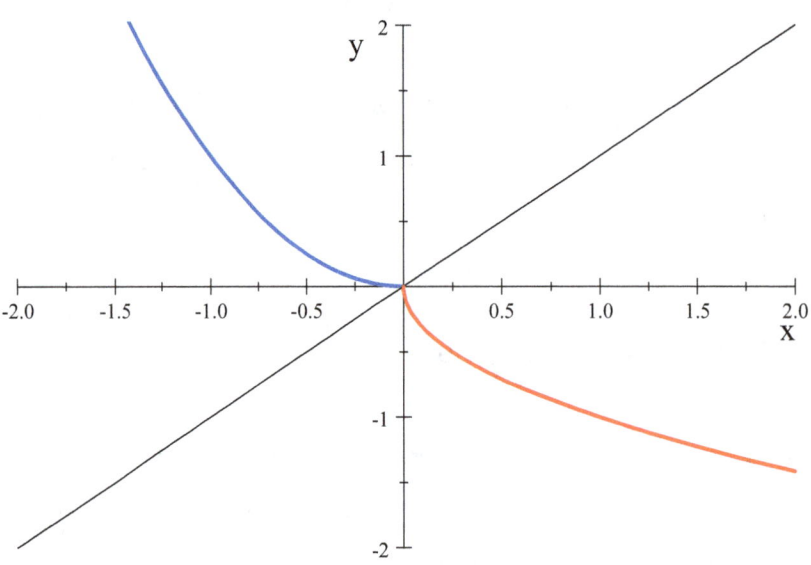

In the picture we can observe the cut graph of x^2 (in blue) and $-\sqrt{x}$ (in red).

3.5.3 Exponential functions

Exponential functions considering $\mathbb{R} \longrightarrow \mathbb{R}$ are not surjective and then their inverse functions do not exist; their restrictions $\mathbb{R} \longrightarrow \mathbb{R}_+ - \{0\}$ are one to one correspondences and their inverse functions are the logarithms (with the same bases); these are functions $\mathbb{R}_+ - \{0\} \longrightarrow \mathbb{R}$

When the base is greater than 1 we have:

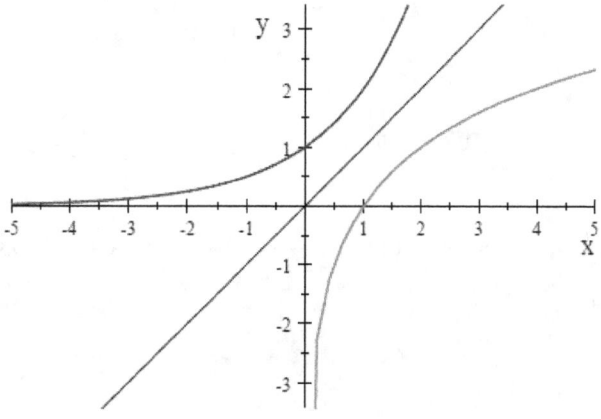

When the base is between 0 and 1:

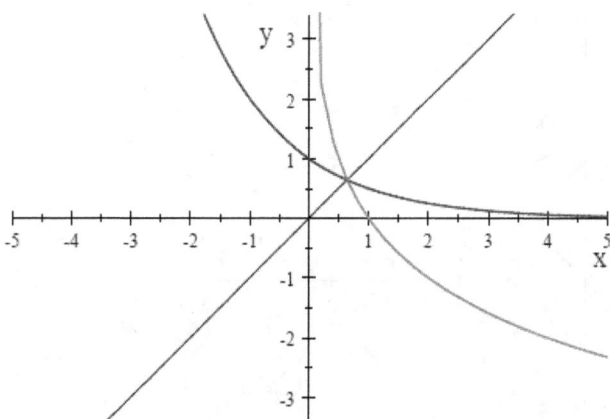

3.5.4 Function $\sin x$

Trigonometric functions are not one to one correspondences unless we restrict their domains and ranges; let's now consider

$$f(x) = \sin x : \left[-\frac{\pi}{2}; \frac{\pi}{2}\right] \longrightarrow [-1; 1]$$

As we saw above now we have a one to one correspondence and we call its inverse function

$$\arcsin x : [-1; 1] \longrightarrow \left[-\frac{\pi}{2}; \frac{\pi}{2}\right]$$

3.5.5 Function $\cos x$

We consider $f(x) = \cos x : [0; \pi] \longrightarrow [-1; 1]$ and we call its inverse function $\arccos x : [-1; 1] \longrightarrow [0; \pi]$

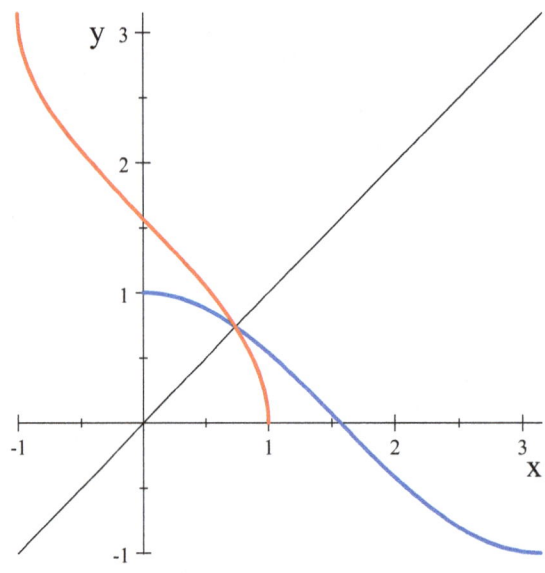

3.5.6 Trigonometric tangent

We consider $f(x) = \tan x : \left]-\frac{\pi}{2}; \frac{\pi}{2}\right[\longrightarrow \mathbb{R}$;

its inverse is called $\arctan x : \mathbb{R} \longrightarrow \left]-\frac{\pi}{2}; \frac{\pi}{2}\right[$. Its graph compared with the one of the tangent is:

$\tan x$

It is important to notice that the cut graph of the tangent is between the vertical lines $x = -\frac{\pi}{2}$ and $x = \frac{\pi}{2}$ while the graph of its inverse function is between the horizontal lines $y = -\frac{\pi}{2}$ and $y = \frac{\pi}{2}$.

Moreover we can say that

$$\lim_{x \to -\infty} \arctan x = -\frac{\pi}{2} \qquad \lim_{x \to +\infty} \arctan x = \frac{\pi}{2}$$

3.5.7 Particular functions

The functions $y = x$, $y = -x$ and $y = \frac{1}{x}$ are the inverse functions of themselves (try if you don't trust me) and indeed their graphs are symmetric with respect to the line $y = x$.

Remark 3.2

1. *The functions presented in this chapter that are one to one correspondences, apart from $y = \frac{1}{x}$, are striclty increasing or decreasing.*

2. *When the tangent line to the graph of f in a point is horizontal the tangent line to the graph of f^{-1} in the corresponding point is vertical.*

At the moment we don't have the right tools to show why these two events occurs; later in this book we will analyze these situations.

3.5.8 How to write the algebraic expression of the inverse function

Raughly speaking we can determine the expression of the inverse function:

1. solving the equation for x;

2. switching x and y.

Example 3.7 *Determine the expression of the inverse function* $y = \log\left(5x^3 + 7\right)$

$$y = \log\left(5x^3 + 7\right)$$
$$\updownarrow$$
$$e^y = 5x^3 + 7$$
$$\updownarrow$$
$$\frac{e^y - 7}{5} = x^3$$
$$\updownarrow$$
$$x = \sqrt[3]{\frac{e^y - 7}{5}}$$

All the above equations describe the same function; the inverse function we are looking for is:

$$y = \sqrt[3]{\frac{e^x - 7}{5}}$$

An economic application

Economists consider the demand of a certain good (quantity that consumers ask to the market) a decreasing function of the price: if the price is very high, very few consumers will buy that good, if the price is low many consumers will ask for it. An example may be the function:

$$q = 100 - 2p$$

here q is the quantity that the market wants to buy and it depends on the price (p). The function is a one to one correspondence and we can write the price as a function of the quantity

$$p = 50 - \frac{1}{2}q$$

The latter equation is the same as the first one written in another way but economists call it inverse demand function. It make sense since if we buy in stock (if we ask for a very big quantity) we can get a better (lower) price.

3.6 Absolute value (modulus)

A very important function is the absolute value denoted by $|x|$; somehow we can think to the absolute value of a number as its value with the positive sign; so it holds: $|+2| = |-2| = -(-2) = +2$. The formal definition is

$$|x| = \begin{cases} +x & \text{if } x \geq 0 \\ -x & \text{if } x \leq 0 \end{cases}$$

The following picture displays the graph of $f(x) = |x|$ in blue; the thin black line is the function $y = x$. Note that when $x \geq 0$ the graph of $|x|$ overlap x; when x is smaller than 0 the graph of $|x|$ is obtained by a simmetry with respect to the horizontal axis.

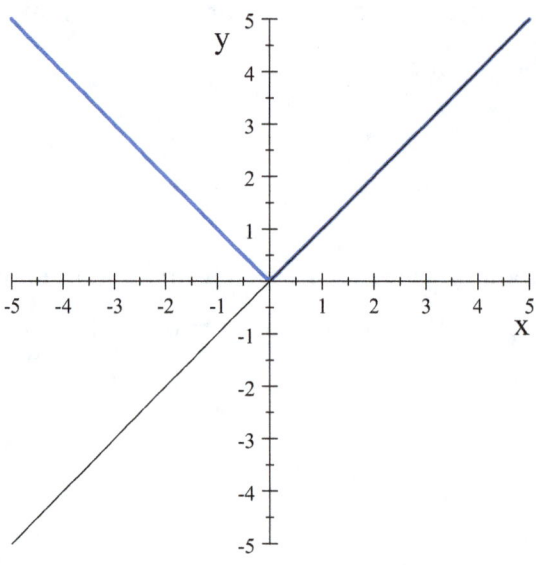

If we know the graph of any function $f(x)$ it is easy to plot $|f(x)|$ by leaving unchanged the graph where $f(x)$ is above the horizontal axis and performing a simmetry with respect of the horizontal axis of the graph when $f(x) \leq 0$.

Example 3.8 *Let us represent $f(x) = |\log x|$*

$$|\log x| = \begin{cases} +\log x & \text{if } \log x \geq 0 \\ -\log x & \text{if } \log x \leq 0 \end{cases} = \begin{cases} +\log x & \text{if } x \geq 1 \\ -\log x & \text{if } 0 < x \leq 1 \end{cases}$$

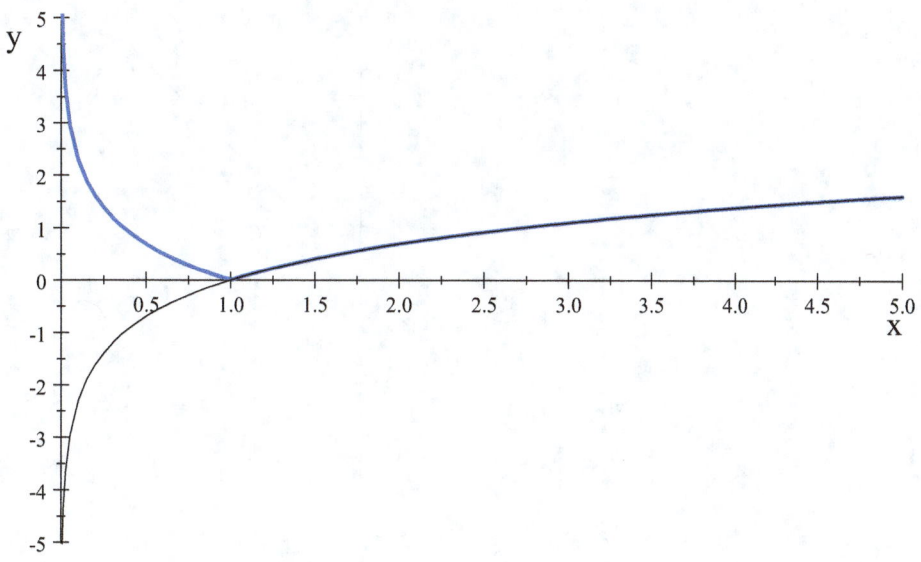

Exercise 3.3 *plot the functions*

$$y = |\sin x|$$
$$y = |x^2 - 3x + 2|$$

3.7 Integer part function (floor)

The integer part function is a very important function in computer science where it is not possible to represent irrational numbers (neither using fixed point nor using floating point notation) since the memory of any machine is finite. The integer part of x is denoted by $\lfloor x \rfloor$ and is the biggest integer number smaller or equal to x.

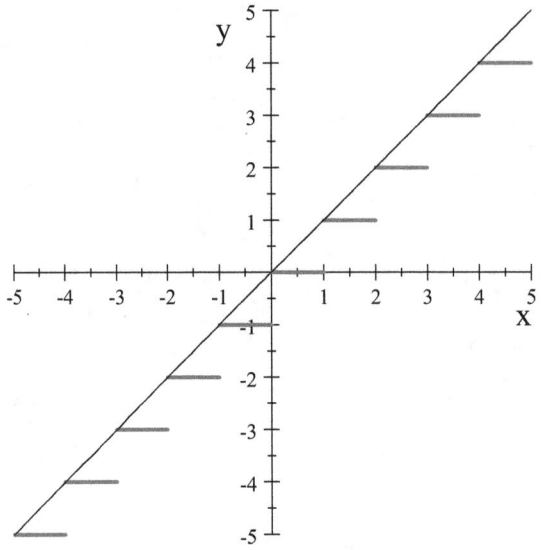

So we have $\lfloor 3 \rfloor = 3$, $\lfloor 3.1 \rfloor = 3$, $\lfloor -3.1 \rfloor = -4$.

Exercise 3.4 *plot the following functions:*

$$y = \lfloor \sin x \rfloor$$
$$y = \lfloor \log_2 x \rfloor$$

Fractional part

The fractional part of x is denoted by $\{x\}$ and we have

$$\{x\} \stackrel{def}{=} x - \lfloor x \rfloor$$

In blue we have its graph; in black we have $y = x$ and $y = \lfloor x \rfloor$:

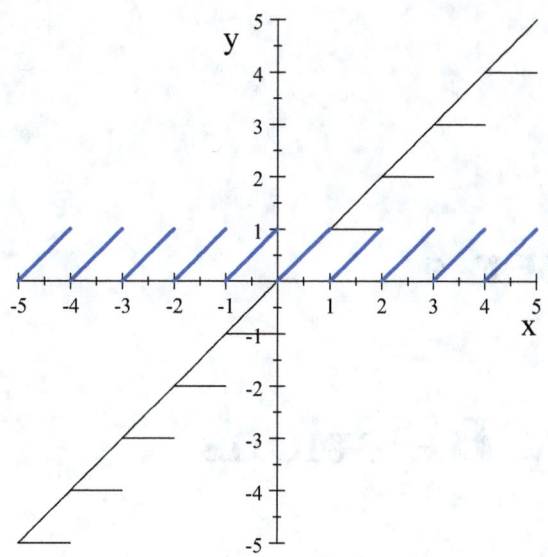

Chapter 4

Binary Relations

4.1 Relations properties

In this section we consider a relation $\mathcal{R}$ from a set A into itself ($\mathcal{R}: A \longrightarrow A$); in this case we say that $\mathcal{R}$ is a relation on A.

Definition 4.1 $\mathcal{R}: A \longrightarrow A$ *is total (or complete or connected) if:*

$$[(a, b \in A) \ and \ (a \neq b)] \implies [(a\mathcal{R}b) \ or \ (b\mathcal{R}a)]$$

or equivalentely:

$$\forall a, b \in A, a \neq b \implies [(a\mathcal{R}b) \ or \ (b\mathcal{R}a)]$$

This property implies that given any two elements we can always "compare" them.

Examples of total relations are: $<$ and $\leq$ between numbers (given any two different numbers we can always understand which one is the smallest).

Definition 4.2 $\mathcal{R}: A \longrightarrow A$ *is reflexive if*

$$\forall a \in A : a\mathcal{R}a$$

The relations $=$, $\leq$ between numbers, $\subseteq$ and $=$ between sets are reflexive; $<$ (between numbers) and $\subset$ (between sets) are not reflexive.

Definition 4.3 $\mathcal{R} : A \longrightarrow A$ *is irreflexive if*

$$\forall a \in A : a \not\mathcal{R} a$$

The relations $=$, $\leq$ between numbers, $\subseteq$ and $=$ between sets are not irreflexive; $<$ (between numbers) and $\subset$ (between sets) are irreflexive

A relation cannot be reflexive and irreflexive at the same time; there are relations that are neither reflexive nor irreflexive.

Definition 4.4 $\mathcal{R} : A \longrightarrow A$ *is symmetric if*

$$a\mathcal{R}b \implies b\mathcal{R}a$$

Examples of symmetric relations are "x is y's brother" between men, $=$ between numbers and between sets, "parallel" between lines; $\leq$, $<$, $\subset$, $\subseteq$ are not symmetric.

Definition 4.5 $\mathcal{R} : A \longrightarrow A$ *is antisymmetric if*

$$[(a\mathcal{R}b) \text{ and } (b\mathcal{R}a)] \implies a = b$$

Examples of antisymmetric relations are $=$ between numbers and between sets, $\leq$, $<$, $\subset$, $\subseteq$. Note that $<$ and $\subset$ are antisymmetric because the assumption is always false: $[(a < b) \text{ and } (b < a)]$ cannot be true. A relation can be symmetric and antisymmetric at the same time (like $=$) and we have relations neither symmetric nor antisymmetric.

Definition 4.6 $\mathcal{R}: A \longrightarrow A$ *is transitive if*

$$[(a\mathcal{R}b) \text{ and } (b\mathcal{R}c)] \implies (a\mathcal{R}c)$$

examples of transitive relations are $=$ between numbers and between sets, $\leq$, $<$, $\subset$, $\subseteq$, "parallel" between lines.

4.2 Equivalence relations

Definition 4.7 *An equivalence relation is a relation:*

reflexive, symmetric and transitive.

Given an equivalence relation $\mathcal{R}$, if $a\mathcal{R}b$ we say tha a and b are equivalent.

Let $\mathcal{R}$ be an equivalence then we can define the equivalence sets (or equivalence classes):

Definition 4.8 *Let $\mathcal{R}: A \longrightarrow A$ be an equivalence, and $a \in A$; the equivalence set of a is the subset of A whose elements are in relation with (equivalent to) a:*

$$[a]_\mathcal{R} \stackrel{def}{=} \{x \in A : a\mathcal{R}x\}$$

The set containing all the equivalence sets is called quotient set.

4.2.1 Properties of the equivalence sets

1. An equivalence set is never empty (at least it contains a):

    $$[a]_\mathcal{R} \neq \varnothing$$

2. Any two equivalence sets either are disjoint or are the same set:

$$\{[a]_\mathcal{R} = [b]_\mathcal{R}\} \ or \ \{[a]_\mathcal{R} \cap [b]_\mathcal{R}\} = \emptyset$$

3. The union of all the equivalence sets is A; that is if $A_1, A_2, ..., A_n$ are the equivalence sets then

$$\bigcup_{k=1}^{n} A_k = A$$

Exercise 4.1 *Let $A := \{a, b, c\}$ then:*

1. build $\mathcal{P}(A)$ ($\mathcal{P}(A)$ is the power set of A that is the set of all subsets of A);

2. consider the relation $\mathcal{N} : \mathcal{P}(A) \longrightarrow \mathcal{P}(A)$:

$$A_i \mathcal{N} A_j \stackrel{def}{\Longleftrightarrow} Card(A_i) = Card(A_j)$$

($Card(X)$ is the number of elements of X) and prove that $\mathcal{N}$ is an equivalence;

3. build the quotient set.

4.2.2 An economic application

A wellknown microeconomic topic is the consumer's choices. The so called consumer's indifference relation among different bundles (sets of goods) available states that two bundles are indifferent if they give to the consumer the same satisfaction. Economists assume that the indifference relation is: reflexive (bundle X gives the same satisfaction as itself), symmetric (if X is indefferent to Y, then Y is indifferent to X) and transitive; it is an equivalence relation and the equivalence sets (sets of bundles giving the same satisfaction) are called indifference curves.

4.3 Order theory

A binary relation is a weak order if it is:

reflexive, antisymmetric and transitive.

If the relation is total then it is a total weak order (partial if it is not total). $\leq$ between numbers is an example of total weak order.

A strict order is a relation that is:

irreflexive, antisymmetric and transitive.

We speak of total strict order or partial strict order if the relation is total or not. The relation $<$ between numbers is a total strict order.

It is important to realize that if an order is total then it is always possible to compare any two elements. One of the most important assumption of the economic theory is that the consumers'preferences are total that is consumers always know what is better for them.

A preorder is a relation:

reflexive and transitive (and it can be total or partial).

Order and preorder are different because of the antisymmetric property; without it we cannot order the elements one by one, but we can order sets (classes) of elements.

Given an order (or a preorder) $\preceq$ on the set A the element $x \in A$ is called maximum (greatest) of A if:

$$a \preceq x \forall a \in A$$

x is called minimum (smallest) of A if:

$$x \preceq a \forall a \in A$$

x is called maximal of A if:

$$\nexists a \in A, a \neq x : x \preceq a$$

x is called minimal of A if:

$$\nexists a \in A, a \neq x : a \preceq x$$

Put it in other words: a maximal is not the greatest but there are no elements in A greater than it.

Exercise 4.2 *Consider the following relations on $\mathbb{Z}$ ($\mathbb{Z}$ is the set of the integer numbers $\mathbb{Z} = \{0, \pm 1, \pm 2, ...\}$) and determine their properties:*

$$x\mathcal{P}y \stackrel{def}{\iff} x^2 = y^2$$
$$x\mathcal{Q}y \stackrel{def}{\iff} x^2 \leq y^2$$
$$x\mathcal{R}y \stackrel{def}{\iff} x^3 \leq y^3$$

Which one is an equivalence? Which is the quotient set? Which one is a preorder? Which one is an order? What changes if we consider the same relations on $\mathbb{N}$ ($\mathbb{N} = \{1, 2, 3, ...\}$)?

4.3.1 An economic Application

In the consumer theory we consider the preference relation among several bundles: bundle X is preferred to bundle Y if X gives to the consumer more satisfaction than Y. Economists assume that this relation is reflexive (a bundle is good at least as much as itself), transitive (...) and complete (the consumer can always compare different alternatives); so it is a total preorder.

Keep in mind that pure mathematicians do not consider preorder very important, but in the economic and financial fields preorder relations are quite common.

4.4 Suggested exercises

1. Analyze the following relations; in the case of equivalence specify the equivalence sets (classes), in the case of pre-order specify wich elements are in the same set, in the case of order or preorder specify greatest and smallest elements (if any), maximals, minimals.

   "is before" according to the alphabetic (lexicographic) order of surnames among a set of individuals;

   "is before" according to the alphabetic (lexicographic) order of surnames-names among a set of individuals;

   "started the university the same year of" among individuals studying in Siena;

   "started the university before or the same year of" among individuals studying in Siena;

2. let $A := \left\{\frac{\pi}{6}, \frac{\pi}{4}, \frac{\pi}{3}, \frac{\pi}{2}\right\}$; analyze the relation:

   $$\mathcal{R} := \left\{\left(\frac{\pi}{6}, \frac{\pi}{6}\right), \left(\frac{\pi}{4}, \frac{\pi}{4}\right), \left(\frac{\pi}{3}, \frac{\pi}{3}\right), \left(\frac{\pi}{2}, \frac{\pi}{2}\right), \left(\frac{\pi}{6}, \frac{\pi}{3}\right), \left(\frac{\pi}{6}, \frac{\pi}{2}\right), \left(\frac{\pi}{4}, \frac{\pi}{2}\right)\right\}$$

   What about the relation:

   $$\overline{\mathcal{R}} := \left\{\left(\frac{\pi}{6}, \frac{\pi}{3}\right), \left(\frac{\pi}{6}, \frac{\pi}{2}\right), \left(\frac{\pi}{4}, \frac{\pi}{2}\right)\right\} \quad ?$$

3. Analyze the following relations:

   $$\mathcal{R}_1 : \mathbb{N} \longrightarrow \mathbb{N}, m\mathcal{R}_1 n \stackrel{def}{\Longleftrightarrow} GCD(m, n) = 1$$

   $GCD(m, n)$ is the greatest common divisor between m and n.

   $$A := \{1, 2, ..., 12\}$$
   $$\mathcal{R}_2 : A \longrightarrow A, m\mathcal{R}_2 n \stackrel{def}{\Longleftrightarrow} \text{"}m \text{ is a divisor (multiple) of } n\text{"}$$

consider the same relation on the sets $B := \{1, 2, ..., 12\} \cup \{60\}$ and $C := \{2^k\}_{k \in \mathbb{N}}$.

$$\mathcal{R}_3 : \mathbb{N} \longrightarrow \mathbb{N}, m\mathcal{R}_3 n \overset{def}{\Longleftrightarrow} \text{"}m \text{ is a divisor (multiple) of } n\text{"}$$

can you plot this relation?

4. Let S be a set; analyze the following relations $\mathcal{P}(S) \longrightarrow \mathcal{P}(S)$:

$$A\mathcal{R}B \overset{def}{\Longleftrightarrow} Card(A) = Card(B)$$
$$A\mathcal{R}B \overset{def}{\Longleftrightarrow} Card(A) \leq Card(B)$$
$$A\mathcal{R}B \overset{def}{\Longleftrightarrow} A \subseteq B$$
$$A\mathcal{R}B \overset{def}{\Longleftrightarrow} A \cap B = \emptyset$$
$$A\mathcal{R}B \overset{def}{\Longleftrightarrow} A = \complement(B)$$

($\complement$ denotes the complement).

Suppose that $S := \{a, b, c\}$; if the relation is an equivalence identify the equivalence class, if the relation is an order or a preorder determine minimum and maximum (if any) or minimal and maximal elements.

5. Let $\mathbb{Z}^* \overset{def}{=} \mathbb{Z} - \{0\}$; consider the relation $\mathcal{R} : \mathbb{Z}^* \longrightarrow \mathbb{Z}^*$:

$$x\mathcal{R}y \overset{def}{\Longleftrightarrow} xy > 0$$

Prove that it is an equivalence, identify and plot the equivalence sets.

6. Let $x, y \in \mathbb{R}$; analyze the following relations:

$$x\mathcal{R}y \overset{def}{\Longleftrightarrow} x = \pm y$$
$$x\mathcal{R}y \overset{def}{\Longleftrightarrow} x^2 + y^2 + 1 = 0$$
$$x\mathcal{R}y \overset{def}{\Longleftrightarrow} (x - y) = 1$$
$$x\mathcal{R}y \overset{def}{\Longleftrightarrow} |x - y| = 1$$
$$x\mathcal{R}y \overset{def}{\Longleftrightarrow} (x - y)^2 = 1$$
$$x\mathcal{R}y \overset{def}{\Longleftrightarrow} \sin^2 x = \cos^2 y$$

4.4.1 Solved excercises

1. Let $A := \{0, 1, 2, ..., 24\}$, consider the relation on A:

   $$a\mathcal{R}b \overset{def}{\iff} \exists k \in \mathbb{Z} : (a-b) = k \cdot 12$$

   This relation is reflexive since:

   $$\forall x \in A, (x-x) = 0 = 0 \cdot 12$$

   and $0 \in \mathbb{Z}$.

   It is also symmetric since:

   $$\forall x, y \in A, [(x-y) = k \cdot 12] \implies [(y-x) = -k \cdot 12]$$

   and if $k \in \mathbb{Z}$ then $-k \in \mathbb{Z}$.

   It is transitive as well since:

   $$\forall x, y, z \in A \left. \begin{array}{l} (x-y) = k_1 \cdot 12 \\ (y-z) = k_2 \cdot 12 \end{array} \right\} \implies (x-z) = (k_1 + k_2) \cdot 12$$

   clearly if k_1 and $k_2 \in \mathbb{Z}$ also $(k_1 + k_2) \in \mathbb{Z}$.

   This is an equivalence and the equivalence sets are:

   $\{0, 12, 24\}, \{1, 13\}, \{2, 14\}, ..., \{11, 23\}$ this is the widely used convention that identifies the 1PM with 13 o'clock and so on.

2. Let $A = [0; 2\pi[$, let's consider the relation:

   $$x\mathcal{R}y \overset{def}{\iff} \left[(\sin x)^2 + (\cos y)^2\right] = 1$$

   The relation is reflexive that is $\forall x, x\mathcal{R}x$ because of the fundamental identity:

   $$(\sin x)^2 + (\cos x)^2 = 1$$

$\mathcal{R}$ is symmetric as well; to prove it we need to perform the substitutions:

$$(\sin x)^2 = 1 - (\cos x)^2$$
$$(\cos y)^2 = 1 - (\sin y)^2$$

it follows:

$$x\mathcal{R}y$$
$$\Updownarrow$$
$$(\sin x)^2 + (\cos y)^2 = 1$$
$$\Updownarrow$$
$$1 - (\cos x)^2 + 1 - (\sin y)^2 = 1$$
$$\Updownarrow$$
$$(\sin y)^2 + (\cos x)^2 = 1$$
$$\Updownarrow$$
$$y\mathcal{R}x$$

The relation is transitive too since:

$$\left.\begin{array}{c} x\mathcal{R}y \\ y\mathcal{R}z \end{array}\right\} \overset{def}{\Longleftrightarrow} \left\{\begin{array}{c} (\sin x)^2 + (\cos y)^2 = 1 \\ (\sin y)^2 + (\cos z)^2 = 1 \end{array}\right\}$$

adding the two equalities side by side we get

$$(\sin x)^2 + \left[(\cos y)^2 + (\sin y)^2\right] + (\cos z)^2 = 1 + 1$$
$$(\sin x)^2 + 1 + (\cos z)^2 = 1 + 1$$
$$(\sin x)^2 + (\cos z)^2 = 1$$
$$\Updownarrow$$
$$x\mathcal{R}z$$

So the relation is an equivalence and the indifference classes are:

$$[\alpha]_\mathcal{R} = \{\alpha, \pi - \alpha, \pi + \alpha, 2\pi - \alpha\}$$

Please consider the graph below:

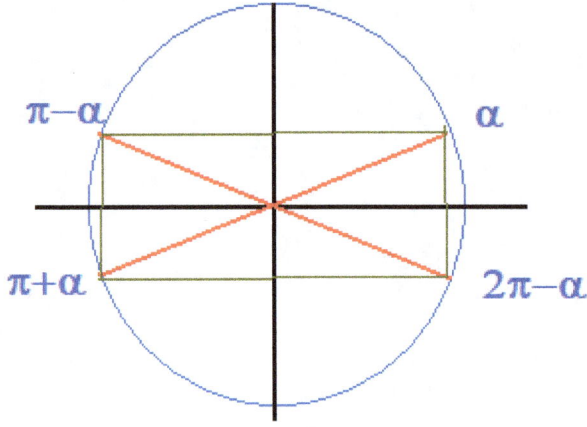

3. Pareto Order

Let us consider the relation $\preceq_P \colon \mathbb{R}^2 \longrightarrow \mathbb{R}^2$

$$\mathbf{x} = \begin{pmatrix} x_1 \\ x_2 \end{pmatrix} \preceq_P \mathbf{y} = \begin{pmatrix} y_1 \\ y_2 \end{pmatrix} \stackrel{def}{\Longleftrightarrow} \begin{cases} x_1 \leq y_1 \\ x_2 \leq y_2 \end{cases}$$

It is easy to verify that the relation is reflexive, antisymmetric and transitive; so it is a weak partial (some elements cannot be compared) order. The following picture helps us to better understand this relation (here we are considering only positive values for the variables):

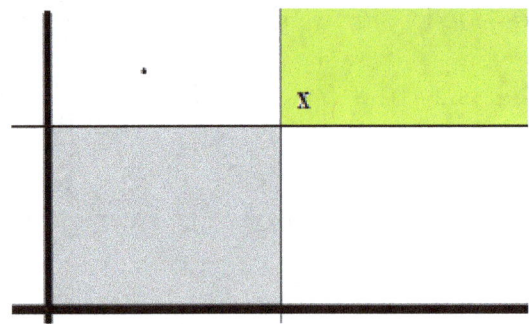

If $x \preceq_P y$ we say that y dominates x and that x is dominated by y. The points in the grey area are the ones dominated by x, the ones in the green area are dominating x. The points in the white areas are not comparable to x.

It is important to understand (using the pictures) which one are the greatest-smallest (minimal or maximal) elements in the following sets:

(a)

$$A := \{(x,y) \in \mathbb{R}^2 : x \geq 0, \ y \geq 0, \ x+y \leq 0\}$$

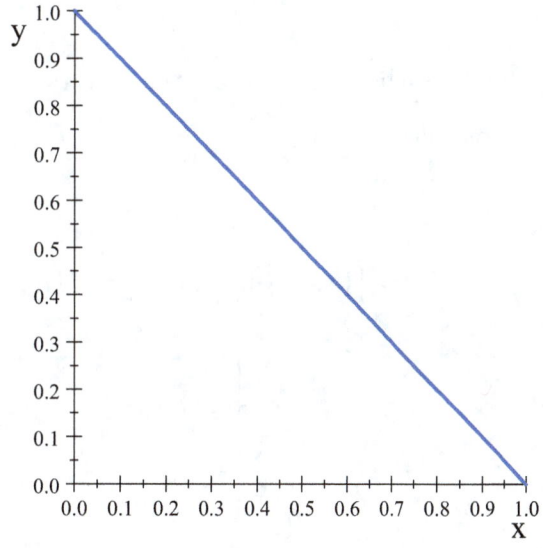

$(0;0)$ is the smallest, all the points on the segment joining $(1;0)$ to $(0;1)$ are maximal (not maximum).

(b)

$$B := \{(x,y) \in \mathbb{R}^2 : x \geq 0, \ y \geq 0, \ x^2 + y^2 \leq 1\}$$

73

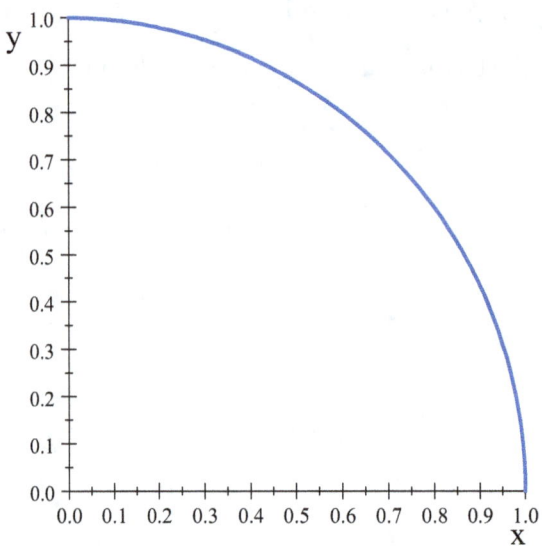

$(0;0)$ is the smallest, all the points on the circle boundary are maximal (not maximum).

(c)

$$C := \{(x,y) \in \mathbb{R}^2 : 0 \leq x \leq 1,\ 0 \leq y \leq 1\}$$

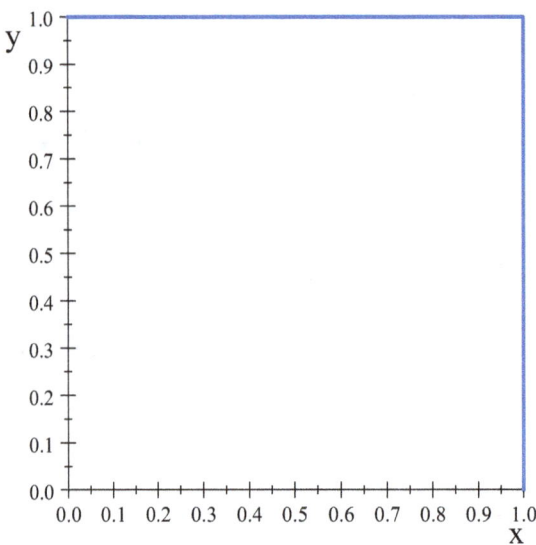

$(0;0)$ is the smallest, $(1;1)$ is the greatest.

Chapter 5

Numerical sets

5.1 Natural numbers

The set of the natural numbers, denoted by the symbol $\mathbb{N}$, is the set containing the positive integer numbers $1, 2, 3, ...$; we introduce the natural numbers using a set of axioms stated by Giuseppe Peano (Italian mathematician born in Cuneo, close to Torino). In his first formulation Peano didn't include the 0 in the set of the natural numbers; later he included also the 0. Nowdays most of the mathematicians does include the 0 among the natural numbers, but, apart from a few examples, in this book we do not include it.

Peano's axioms

$\mathcal{A}1$: $1 \in \mathbb{N}$;

$\mathcal{A}2$: there exists the successor function $\sigma : \mathbb{N} \longrightarrow \mathbb{N} - \{1\}$;

$\mathcal{A}3$: σ is surjective (this means that every natural number different by 1 is the successor of some number);

$\mathcal{A}4$: σ is injective (different numbers have different successors);

$\mathcal{A}5$: induction principle

Let $T \subseteq \mathbb{N}$ then:

$$\left.\begin{array}{r} 1 \in T \\ [n \in T] \implies [\sigma(n) \in T] \end{array}\right\} \implies [T = \mathbb{N}]$$

Remark 5.1 *The output of σ for any natural number n is its successor, that is $\sigma(n) = n + 1$; the induction principle states that, if $1 \in T$, and if, the belonging of any number to T implies the belonging of its successor to T, then T is equal to $\mathbb{N}$. Hence: $1 \in T$, and this implies that $2 \in T$; hence also the successor of 2 (3) belongs to T, and so on in a neverending process.*

Applications

1. Prove that:

$$1 + 2 + \ldots + n = \frac{n(n+1)}{2}$$

Let T be the subset of $\mathbb{N}$ satisfying the above equality; clearly $1 \in T$; let's prove that if the equality works for a particular number n then it works for its successor $\sigma(n) = n + 1$:

$$\sum_{k=1}^{n} k = 1 + 2 + \ldots + n = \frac{n(n+1)}{2}$$

$$\Downarrow$$

$$1 + 2 + \ldots + n + (n+1) = \frac{n(n+1)}{2} + (n+1)$$

$$1 + 2 + \ldots + n + (n+1) = \frac{n(n+1) + 2(n+1)}{2}$$

$$1 + 2 + \ldots + n + (n+1) = \frac{(n+1)(n+2)}{2}$$

now the proof is complete and we can conclude that $T = \mathbb{N}$; this means that the equality works for every natural number.

2. Sum of a geometric sequence. Let $q \neq 0$, $q \neq 1$; in this example we include the 0 in $\mathbb{N}$ that then:

$$\sum_{k=0}^{n} q^k = q^0 + q^1 + ... + q^n = \frac{1 - q^{n+1}}{1 - q}$$

Clearly the equality works with $n = 0$: $q^0 = \frac{1-q}{1-q} = 1$. Now we should proove that if the equality works for n then it works for $\sigma(n)$:

$$q^0 + q^1 + ... + q^n = \frac{1 - q^{n+1}}{1 - q}$$

$$\Downarrow$$

$$(q^0 + q^1 + ... + q^n) + q^{n+1} = \frac{1 - q^{n+1}}{1 - q} + q^{n+1}$$

$$q^0 + q^1 + ... + q^n + q^{n+1} = \frac{1 - q^{n+1} + q^{n+1} - q^{n+2}}{1 - q} = \frac{1 - q^{n+2}}{1 - q}$$

and so the equality works $\forall n \in \mathbb{N}$.

At the beginning of this example we assume that $q \neq 0$ and $q \neq 1$; we cannot assume $q = 0$ since 0^0 is not defined, and q cannot be equal to one because the denominator cannot be equal to zero. What happens if we remove these assumptions?

The answer is not so difficult:

if $q = 0$ then we can start the sum with $k = 1$ and $\sum_{k=1}^{n} q^k = 0 + 0 + ... + 0 = 0$;

if $q = 1$ then $\sum_{k=0}^{n} q^k = 1 + 1 + ... + 1 = n + 1$.

Here I want to suggest to the reader to remember this example since it is widely used in financial and economic application.

Moreover since $\mathbb{N}$ is a neverending set it make sense to wonder what happens when n becomes bigger and bigger, so we can consider the expression:

$$\lim_{n \longrightarrow +\infty} \sum_{k=1}^{n} q^k = \lim_{n \longrightarrow +\infty} \frac{1 - q^{n+1}}{1 - q}$$

Here we focus our attention on the case $0 < q < 1$; in the first chapter we observed that when $0 < a < 1$

$$\lim_{n \longrightarrow +\infty} a^n = 0$$

and so it follows that

$$\lim_{n \longrightarrow +\infty} \sum_{k=1}^{n} q^k = \lim_{n \longrightarrow +\infty} \frac{1 - \boxed{q^{n+1}}}{1 - q} = \frac{1}{1 - q}$$

since the expression in the box tends to 0.

3. Prove that the sum of the first n odd numbers is equal to n^2:

$$\sum_{k=1}^{n} (2k - 1) = 1 + 3 + 5 + ... + (2n - 1) = n^2$$

The proof is left to the reader.

5.2 Integer numbers

The set of the integer numbers is denoted by $\mathbb{Z}$ and contains all the integer numbers (positive and negative) including the 0. So we can write

$$\mathbb{Z} = \{0, \pm 1, \pm 2, ...\}$$

Probally the origin of negative numbers has a financial root with positive numbers denoting credits, negative numbers denoting liabilities and 0 denoting a balanced budget.

We formally introduce $\mathbb{Z}$ by considering the following binary relation

$$\mathcal{D} : [\mathbb{N} \cup \{0\}] \times [\mathbb{N} \cup \{0\}] \longrightarrow [\mathbb{N} \cup \{0\}] \times [\mathbb{N} \cup \{0\}]$$

$$(x,y)\,\mathcal{D}\,(a,b) \overset{def}{\Longleftrightarrow} [x+b = y+a]$$

It is easy to proove that it is an equivalence since:

$$\forall\,(x,y),\,[x+y = y+x] \Longleftrightarrow (x,y)\,\mathcal{D}\,(x,y)$$

hence the relation is reflexive;

$$\forall\,(x,y),(a,b)$$

$$[x+b = y+a] \Longrightarrow [a+y = b+x]$$

$$\Updownarrow$$

$$(x,y)\,\mathcal{D}\,(a,b) \Longrightarrow (a,b)\,\mathcal{D}\,(x,y)$$

hence the relation is symmetric;

$$\forall\,(x,y),(a,b),(m,q)$$

$$\left.\begin{array}{ll} E_1 & x+b = y+a \\ E_2 & a+q = b+m \end{array}\right\} \overset{(E_1+E_2)}{\Longrightarrow} x+q = y+m$$

$$\Updownarrow$$

$$\left.\begin{array}{l} (x,y)\,\mathcal{D}\,(a,b) \\ (a,b)\,\mathcal{D}\,(m,q) \end{array}\right\} \Longrightarrow (x,y)\,\mathcal{D}\,(m,q)$$

and so the relation is transitive.

The equivalence sets are:

$$[(0,0)]_\mathcal{D} = \{(0,0),(1,1),(2,2),...\}$$

$$[(0,1)]_\mathcal{D} = \{(0,1),(1,2),(2,3),...\}$$

$$[(1,0)]_\mathcal{D} = \{(1,0),(2,1),(3,2),...\}$$

The quotient set is called $\mathbb{Z}$.

The trick to introduce this set is by identifying the equivalence set $[(0,0)]_\mathcal{D}$ with the integer number 0 since $0 - 0 = 0$, $1 - 1 = 0$ and so on. The integer number -1 is the equivalence set $[(0,1)]_\mathcal{D}$ since $0 - 1 = 1 - 2 = 2 - 3... = -1$. So every integer number is represented in the quotient.

5.3 Rational numbers

Let's denote the set of the non-zero integer numbers using the symbol $\mathbb{Z}^*$ ($\stackrel{def}{=} \mathbb{Z} - \{0\}$) and let's consider the following equivalence:

$$\mathcal{D} : [\mathbb{Z} \times \mathbb{Z}^*] \to [\mathbb{Z} \times \mathbb{Z}^*]$$

$$[(a,b)\, \mathcal{D}\, (m,q)] \stackrel{def}{\iff} [a \cdot q = b \cdot m]$$

The set of the rational number, denoted by $\mathbb{Q}$ is the quotient set of this relation.

Remark 5.2 *First of all keep in mind that the symbol $\mathbb{Q}$ stands for quotient. The equivalence set of the pair $(a;b)$ identifies the rational number a/b and this highlight why the second element of each pair cannot be 0; consider, as an example, the equivalence set*

$$[(1,2)]_\mathcal{D} = \{(1,2), (2,4), (3,6) ...\}$$

it identifies the fraction $\frac{1}{2} = \frac{2}{4} = \frac{3}{6} = ...$

5.3.1 Is the rational set enough?

Since a long long long time ago humankind was trying to set a one to one correspondence between the points on a line and a set of numbers. Does $\mathbb{Q}$ achieve this goal?

Unfortunately it doesn't. About 500 years before Jesus Christ Pitagora realized this problem; according to his theorem the hypotenuse of a right triangle whose other two sides are 1 is $\sqrt{2}$ and clearly $\sqrt{2}$ can be represented on a line:

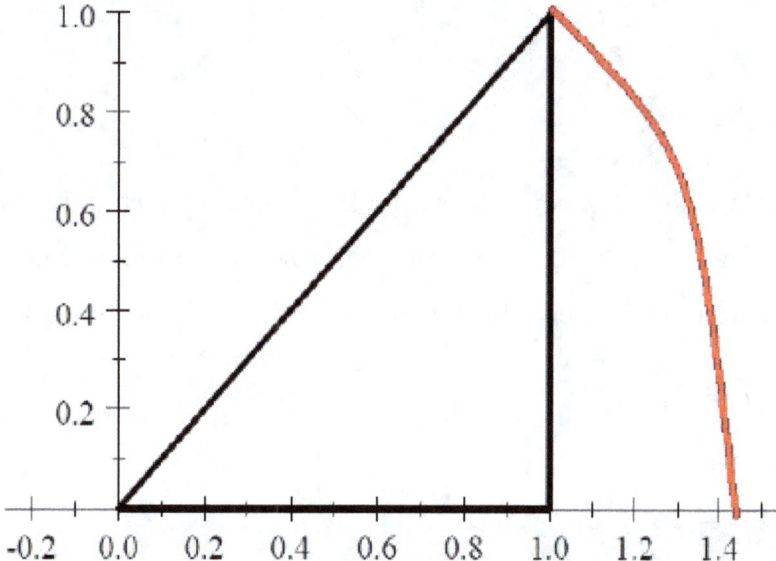

Proposition 5.1 *there is no fraction $\frac{m}{q}$ such that*

$$\frac{m}{q} = \sqrt{2}$$

Proof. This proof is an useful application of the contradiction principle

$$[A \implies B] \iff [not B \implies not A]$$

The goal now is to prove that given any fraction $\frac{m}{q}$ then $\frac{m}{q} \neq \sqrt{2}$.

Let's suppose there exists a fraction $\frac{m}{q}$ (where, without loss of generality, we assume that m and q are positive and have no common divisor aside from 1) such that:

$$\frac{m}{q} = \sqrt{2}$$

$$\Downarrow$$

$$m = \sqrt{2} \cdot q$$

$$\Downarrow$$

$$m^2 = 2q^2$$

The right side of the equality is an even number; the equality can hold only if m^2 is an even number; but if m^2 is even then also m is even. It follows that we can write $m = 2a$ where a is a positive integer number and by substitution we get:

$$4a^2 = 2q^2$$

$$\Downarrow$$

$$2a^2 = q^2$$

Last equality implies that also q is an even number (like m) but now we have a critical contradiction (we assumed that m and k have no common divisor aside from 1). This concludes the proof. ∎

5.4 Real numbers

In order to obtain the one to one correspondence between the points on a line and a numerical set we need a set "wider" than $\mathbb{Q}$; we call this set the set of the real numbers and we denote it by the symbol $\mathbb{R}$. We use an axiomatic definition:

Definition 5.1 *$\mathbb{R}$ is a set with two operators, $+$ (sum) and $\cdot$ (product), and a total weak order $\leq$. By assumption $\mathbb{R}$ has some properties:*

$\mathcal{A}1$ $+$ *is:*

1. *commutative i.e.* $\forall a, b \in \mathbb{R}$ *it holds:*

$$a + b = b + a$$

2. *associative i.e.* $\forall a, b, c \in \mathbb{R}$ *it holds:*

$$(a + b) + c = a + (b + c)$$

3. *existence of the identity element i.e. there exists an element of* $\mathbb{R}$, *denoted by* 0 *such that* $\forall a \in \mathbb{R}$:

$$a + 0 = a$$

4. *existence of the negative (of every element):* $\forall a \in \mathbb{R}$ *there exists an element of* $\mathbb{R}$ *denoted by* $-a$ *such that:*

$$-a + a = 0$$

$\mathcal{A}2$ $\cdot$ *is:*

1. *commutative i.e.* $\forall a, b \in \mathbb{R}$ *it holds:*

$$a \cdot b = b \cdot a$$

2. *associative i.e.* $\forall a, b, c \in \mathbb{R}$ *it holds:*

$$(a \cdot b) \cdot c = a \cdot (b \cdot c)$$

3. *existence of the identity element i.e. there exists an element of* $\mathbb{R}$, *denoted by* 1 *such that* $\forall a \in \mathbb{R}$:

$$a \cdot 1 = a$$

4. *existence of the inverse (reciprocal) of every element different by 0:* $\forall a \neq 0 \in \mathbb{R}$ there exists an element of $\mathbb{R}$ denoted by $\frac{1}{a}$ such that:

$$\frac{1}{a} \cdot a = 1$$

5. *the product is distributive with respect to the sum:* $\forall a, b, c \in \mathbb{R}$ it holds

$$a \cdot (b + c) = a \cdot b + a \cdot c$$

A3 *the total weak order $\leq$ satisfies the properties:*

1. $\forall a, b, x \in \mathbb{R}$ *it holds:*

$$[a \leq b] \iff [a + x \leq b + x]$$

2. $\forall a, b, x \in \mathbb{R}$, $x > 0$ *it holds:*

$$[a \leq b] \iff [a \cdot x \leq b \cdot x]$$

A4 *Completeness axiom (or Dedekind's axiom) Let $(A; B)$ be a partition[1] of $\mathbb{R}$ such that $\forall a \in A$, $\forall b \in B$ it holds $a \leq b$[2], then we have:*

$$\exists! \lambda \in \mathbb{R} : a \leq \lambda \leq b$$

λ is called separating element (or simply separator).

Remark 5.3 *(optional): It is possible to prove that: $\mathbb{R}$ contains a sub-field isomorphic to $\mathbb{Q}$, $\mathbb{Q}$ contains a sub-field isomorphic to $\mathbb{Z}$, this contains a sub-field isomorphic to $\mathbb{N}$. Algebric structures (like fields) are not included in this course. Some authors writes:*

$$\mathbb{N} \subset \mathbb{Z} \subset \mathbb{Q} \subset \mathbb{R}$$

---

[1] to say that A and B are a partition of $\mathbb{R}$ means that A and B are non-empty and

$A \cup B = \mathbb{R}$
$A \cap B = \emptyset$

[2] in such a case we say that (A, B) is a section of $\mathbb{R}$.

This is not quite correct but can be helpful.

Remark 5.4 *The existence of a separating element holds both in $\mathbb{N}$ and in $\mathbb{Z}$ (here it is non unique) bu not in $\mathbb{Q}$. As an example let's consider the following subset of $\mathbb{Q}$:*

$$B := \{q > 0 : q^2 > 2\}$$
$$A := \mathbb{Q} - B$$

Let λ be the unique separating element A and B; it must hold $\lambda^2 = 2$, but such an element does not belong to $\mathbb{Q}$ as we saw in (5.3.1).

Definition 5.2 *The numbers belonging to $\mathbb{R}$ but not belonging to $\mathbb{Q}$ are called irrational numbers.*

Remark 5.5 *The elements of $\mathbb{Q}$, in the decimal representation, after the decimal point have a finite number of digits different by 0, or are recurring (the same sequence of digits is repeated infinity many times; think to the decimal representation of $1/3$). On the countrary irrational numbers have infinity many non recurring digits.*

Since there is a one to one correspondence between $\mathbb{R}$ and the point of a stright line we'll often refer to $\mathbb{R}$ as the real number line (or simply real line), or the real axis.

5.5 Finite and infinite sets

Let S_n be the set of the first n natural numbers; we say that the set A is finite if exists a one to one correspondence between S_n and A and we write $card(A) = n$ (equivalent, widely used notations are $\#(A) = n$ or $2^A = n$). $card(A)$, and the other notations stands for "number of elements of A". If there is no n such that exists a one to one correspondence between S_n and A then we say that A is infinite.

Example 5.1 *Let $A := \{a; b; c\}$, then: $card(A) = 3$.*

Remark 5.6 $\mathbb{N}$ *is an infinite set (because of the induction principle).*

Definition 5.3 *A set A is countable if there exists a one to one correspondence between $\mathbb{N}$ and A.*

Example 5.2 *Let $\mathbb{N}_p$ be the subset of $\mathbb{N}$ containing only the even numbers; it is a countable set. A possible one to one correspondence is:*

$$d : \mathbb{N} \longrightarrow \mathbb{N}_p, \ d(n) = 2n$$

Proposition 5.2 *(no proof is presented)*

1. $\mathbb{Z}$ *is countable;*

2. $\mathbb{Q}$ *is countable;*

3. $\mathbb{R}$ *and the set of the irrational numbers are not countable.*

Chapter 6

Topology of the real axis

The term topology derives from the ancient Greek and means study of the places. Topology is a wide area of mathematics and is not easy to even give an idea of the many concepts covered by this field. In this chapter we deal only with the topics we'll need in the reminder of the book. One of the stepping stone of this branch of the mathematic is the neighborhood; before definying what a neighborhood is we need to introduce the concept of interval.

6.1 Basic definitions

An open interval is a subset of $\mathbb{R}$, denoted by $]a, b[$ or by (a, b); the formal definition is :

$$]a, b[\stackrel{def}{=} \{x \in \mathbb{R} : a < x < b\}$$

A closed interval is a subset of $\mathbb{R}$, denoted by $[a, b]$; the formal definition is :

$$[a, b] \stackrel{def}{=} \{x \in \mathbb{R} : a \leq x \leq b\}$$

Sometimes will use the notation $[a, b[$ or $[a, b)$ to refer to the points $a \leq x < b$ and $]a, b]$ or $(a, b]$ to refer to the points $a < x \leq b$.

In all these cases a and b are called bounds of the interval.

Definition 6.1 *Let $c \in \mathbb{R}$ (c is a point on the real line) a neighborhood of c, denoted as $N(c)$ is an open interval $]a; b[$ such that*

$$a < c < b$$

In the above picture the thin black line is the real axis and the red thick segment is $N(c)$; the empty circles are used to underline that the bounds are not included in the interval (we'll use full circles when they belong to the interval).

Definition 6.2 *(equivalent to the previous one) Let $c \in \mathbb{R}$ a neighborhood of c, is an interval*

$$]c - \epsilon_1, c + \epsilon_2[$$

where $\epsilon_i > 0$ ($i = 1, 2$).

Definition 6.3 *A neighborhood centered in c (also called ball with center c), denoted as $N_\epsilon(c)$ is an interval:*

$$]c - \epsilon, c + \epsilon[$$

where $\epsilon > 0$, is called radius of the neighborhood, and c is called center of the neighborhood; we also refer to $N_\epsilon(c)$ as ϵ–neighborhood of c.

Remark 6.1 *Points x belonging to $N_\epsilon(c)$ satisfy the inequality*

$$|x - c| < \epsilon$$

Since the absolute value $|x - c|$ can be thought as the distance[1] between x and c, $N_\epsilon(c)$ is the set of the points whose distance from c is less than ϵ, and writing $x \in N_\epsilon(c)$ we mean that the distance between x and c is less than ϵ.

6.2 Classification of the points of a set

Let $A \subseteq \mathbb{R}$; we have the following definitions:

- c is an **interior** point of A if there exists a neighborhood of c fully contained in A;

- c is an **exterior** point to A if there exists a neighborhood of c containing no points of A;

- c is a **boundary** point of A if every of its neighborhood contains both points beloning to A and points not belonging to it;

- c is a **detached** (lonely I would say) point of A if:

  1. $x \in A$

  2. there exists a neighborhood of c containing no points of A apart from c;

- c is a **limit** (**accumulation**) point of A if every of its neighborhood contains infinite points of A;

---

[1] In mathematics a distance is a function satisfying some specific properties; here we do not analyze these details.

- (optional) c a point of **closure** of A if every of its neighborhood contains elements of A (one, more than one or infinite ones).

The set of the interior points of A is called interior of A (mathematicians are boring people who lack imagination) and is denoted with $\overset{\circ}{A}$ or $Int(A)$.

The set of the exterior points of A is called exterior of A.

The set of the boundary points of A is called boundary of A and is denoted by $\partial(A)$.

The set of the accumulation points is sometimes denoted writing $D(A)$.

The set of the points of closure of A is called closure of A and denoted by $\overline{A}$.

Example 6.1 *Let's consider:* $A := \{1, 2, 3\} \cup [5, 6[$ *here is a graphical representation of A:*

Red points (and the segment) identify the points of A while the blue segments are neighborhoods; please note that the dot close to 5 is full to show that it belongs to A while the dot close to the 6 is empty to show that it doesn't belong to the set.

It follows:

$\overset{\circ}{A} =]5, 6[$;

set of the exterior points of $A = \complement(A) - \{6\}$;

$\partial(A) = \{1, 2, 3, 5, 6\}$

set of detached points: $\{1, 2, 3\}$;

set of limit points $[5; 6]$;

$\overline{A} = A \cup \{6\}$.

Whatever is $A \subseteq \mathbb{R}$ we have:

- an interior point of A always belongs to A:

$$\left[c \in \overset{\circ}{A}\right] \implies [c \in A]$$

- an interior point of A always is a limit point for A;

- a detached point always is a boundary point;

- The closure of A is equal to

  - A joint to the set of its limit points;
  - A joint to its boundary;
  - the union of the limit points of A and its boundary;
  - the union of the limit points of A and its detached points.

Definition 6.4 *A set containing all its boundary is said closed; a set containing no one of its boundary points is said open; consider the following examples:*

$[1; 2]$ is closed;

$]1; 2[$ is open;

$]1; 2]$ is neither open nor closed;

$\mathbb{R}$ and $\varnothing$; are both open and closed.

6.3 Extremes of a set

Let me remind you that in $\mathbb{R}$ we have a total order $\leq$; let's now take into consideration the following sets:

$$A_c := [1, 2]$$
$$A_o := \,]1, 2[$$

In A_c the boundary points (1 and 2) are the smallest and the greatest elements of the set since $1 \leq x \,\forall x \in A_c$ and...; in A_o the same points are not the smallest and the greatest of the set since they do not belong to A; they are called infimum and supremum of the set and denoted by $\inf(A)$ and $\sup(A)$.

Let us introduce these new concepts in a formal way:

Definition 6.5 *x is an upper bound of A if:*

$$x \geq a \forall a \in A$$

x is a lower bound of A if:

$$x \leq a \forall a \in A$$

Definition 6.6 *The smallest among the upper bounds of A is called supremum of A; the greatest among the lower bounds of A is called infimum.*

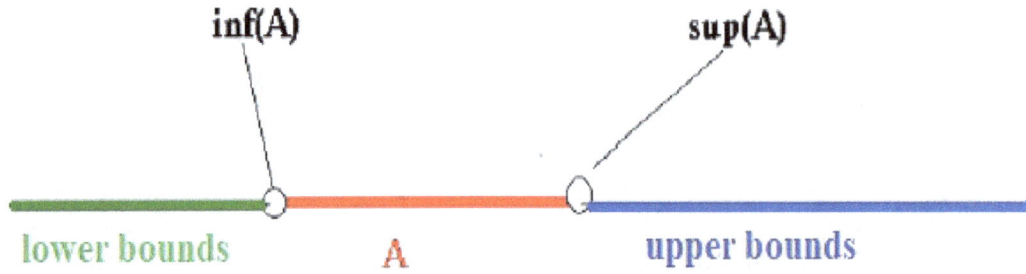

Example 6.2 Let us consider the following sets:

$$A := \left\{\frac{1}{n}, 2 - \frac{1}{n}\right\}_{n \in \mathbb{N}}$$

$$B := \bigcup_{n \in \mathbb{N}} \left[\frac{1}{n}, 2 - \frac{1}{n}\right]$$

$$C := \bigcap_{n \in \mathbb{N}} \left]-\frac{1}{n}, 2 + \frac{1}{n}\right[$$

A) somehow we can represent the set as follows (vertical red segments are the points of A):

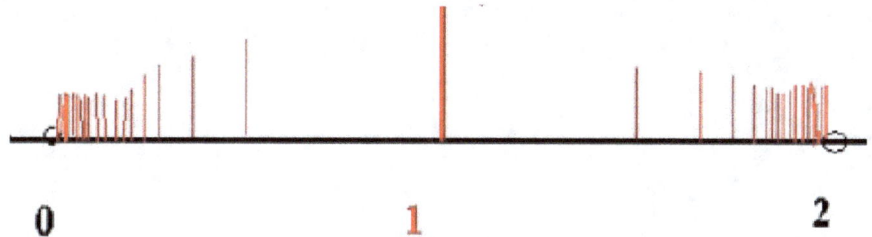

You should realize that the points are closer and closer to each other when n becomes bigger, and they accumulate close to 0 and 2 which don't belong to the set since we'll never reach them (we cannot count up to infinity). Hence:

$\overset{\circ}{A} = \varnothing$;

set of the exterior points of $A = \complement(A) - \{0, 2\}$;

$\partial(A) = A \cup \{0, 2\}$;

set of detached points $= A$;

set of limit points $= \{0, 2\}$;

$\overline{A} = A \cup \{0, 2\}$;

A is neither closed nor open and $\sup(A) = 2$, $\inf(A) = 0$ (they are not greatest and smallest).

B) $B := \bigcup_{n\in\mathbb{N}} \left[\frac{1}{n}, 2 - \frac{1}{n}\right] = \,]0,2[$. The picture below shows somehow the first sets of the sequence forming B:

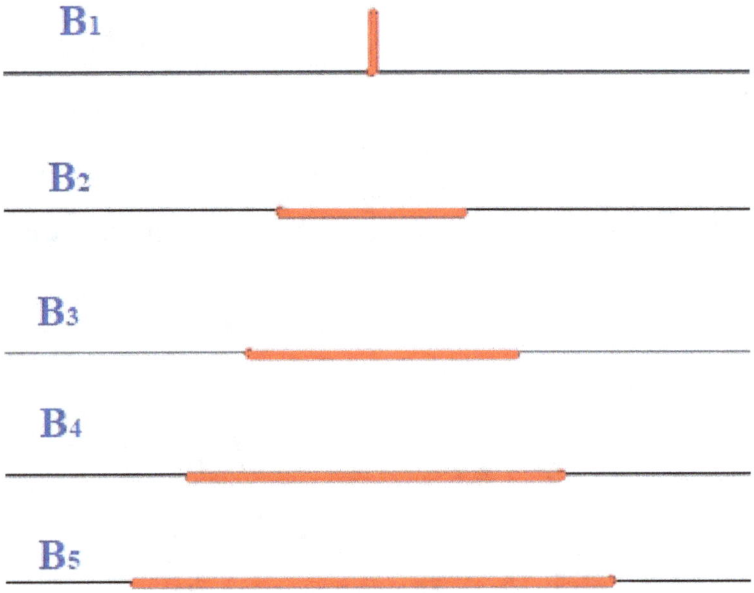

B₁

B₂

B₃

B₄

B₅

Let B_k the interval $\left[\frac{1}{k}, 2 - \frac{1}{k}\right]$, it is important to realize that: $B_k \subset B_{k+1}$ (use numerical examples if needed) and that the intervals become larger and larger but will never reach 0 e 2; for this reason the (infinite but countable) union is $]0; 2[$; moreover:

$\overset{\circ}{B} =]0; 2[$;

set of the exterior points of $B = \complement(B) - \{0, 2\}$;

$\partial(B) = \{0, 2\}$;

set of detached points $= \emptyset$;

set of limit points $= [0, 2]$;

$\overline{B} = B \cup \{0, 2\}$;

$\sup(B) = 2$, $\inf(B) = 0$ *(they are not greatest and smallest).*

B is an open set! This is surprising to my eyes: I was joining closed sets and I got an open set.

C) $C := \bigcap_{n \in \mathbb{N}} \left] -\frac{1}{n}, 2 + -\frac{1}{n} \right[= [0, 2]$. *Let us consider the picture:*

C₁

C₂

C₃

C₄

Let C_k be the interval $\left]-\frac{1}{k}, 2+\frac{1}{k}\right[$; we can see that: $C_{k+1} \subset C_k$ and that the intervals become smaller when k increases but they will never exclude 0 and 2; the intersection is the set $[0; 2]$. Moreover:

$\overset{\circ}{C} =]0; 2[$;

set of the exterior points of $C = \complement(C)$

$\partial(C) = \{0, 2\}$;

set of detached points $= \varnothing$;

set of limit points $= [0, 2]$;

$\overline{C} = C \cup \{0, 2\}$;

$\sup(C) = 2$, $\inf(C) = 0$ *(they are greatest and smallest).*

C is a closed set! This is surprising too: the intersection of (infinite) open sets returns a closed set.

Chapter 7

Limits

In the previous chapter we saw what is a neighborhood of $c \in \mathbb{R}$, (c is a finite number); now we consider a new set, extended $\mathbb{R}$:

$$\overline{\mathbb{R}} \stackrel{def}{=} \mathbb{R} \cup \{+\infty, -\infty\}$$

and some more neighborhood definitions.

Definition 7.1 *A left neighborhood of c (finite) is a set $J^-(c) \stackrel{def}{=}]c - \epsilon; c[$;*

A right neighborhood of c (finite) is a set $J^+(c) \stackrel{def}{=}]c; c + \epsilon[$.

A neighborhood of $+\infty$ is the set

$$N(+\infty) \stackrel{def}{=}]a, +\infty[$$

A neighborhood of $-\infty$ is the set

$$N(-\infty) \stackrel{def}{=}]-\infty, b[$$

Let me remind you that I will use the notation $D(A)$ to denote the set of the limit points of A.

7.1 Universal definition of limit

Using the definitions of neighborhood we can present the following universal definition of limit:

Definition 7.2 *Topological definition of limit: Let $f : A \subseteq \mathbb{R} \longrightarrow \mathbb{R}$; $c \in D(A) \cap \overline{\mathbb{R}}$ then:*

$$\lim_{x \to c} f(x) = l \quad (l \in \overline{\mathbb{R}})$$

if whatever is the neighborhood N of l exists a neighborhood J of c such that for all x belonging to $N(c)$, x different by c then $f(x)$ belongs to $J(c)$; put it in another way:

$$\forall N(l) \, \exists J(c) : x \in J(c), \, x \neq c, \implies f(x) \in N(l)$$

I want to focus the attention on the condition $x \neq c$ which is really significant. When computing a limit what really matters is what happens "close" to c (in a neighborhood of c); not the function value in c (usually we compute the limit when $f(c)$ does not exist and c is a limit point of the domain of f).

Example 7.1 *let's consider $f(x) = \frac{x}{x}$; of course f does not exist if $x = 0$, so 0 does not belong to the domain of f but it is a limit point for D_f. The graph is the following:*

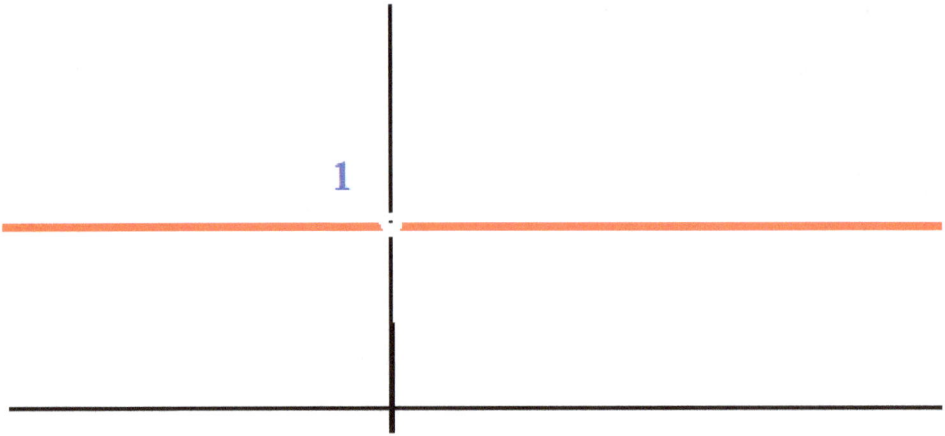

and $\lim_{x \to 0} f(x) = 1$ *since for every x in a neighborhood of 0 (on the horizontal axis) if $x \neq 0$ we have $f(0) = 1$ which belongs to every neighborhood of 1 (on the vertical axes).*

7.2 Subcases of the universal definition

The universal definition has 36 subcases; here I want to present the 15 more relevant cases. I strongly suggest to the reader to represent by himself the graphs presented.

1. $\lim_{x \to +\infty} f(x) = +\infty$

$$\forall N(+\infty) \, \exists J(+\infty) : x \in J(+\infty) \implies f(x) \in N(+\infty)$$

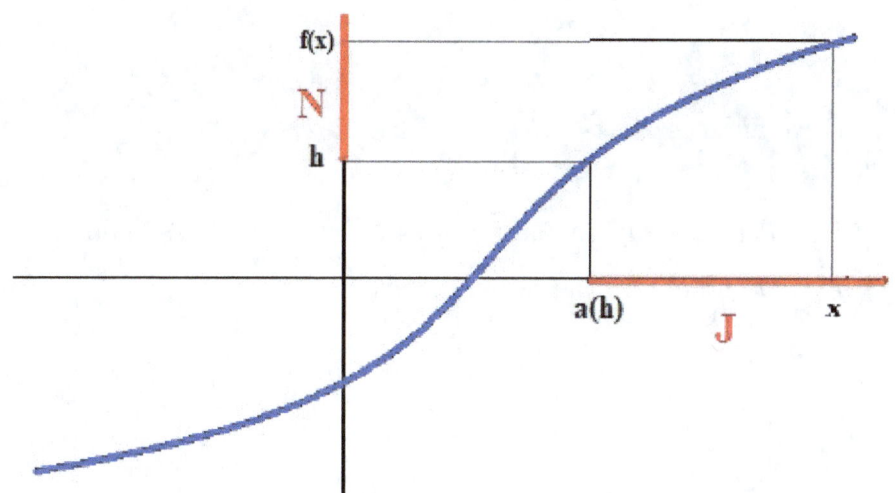

Now please note that: $N(+\infty) =]h, +\infty[$ (here h stands for huge) and $J(+\infty) =]a(h), +\infty[$; we write $a(h)$ to underline that the lower bound of J depends on h (it is function of h) because if we choose a bigger h we need a bigger a. Moreover to write $x \in J(+\infty)$ is the same as to write $x > a(h)$ and to write $f(x) \in N(+\infty)$ is equivalent to write $f(x) > h$. So we have:

$$\forall N(+\infty) \exists J(+\infty) : x \in J(+\infty) \implies f(x) \in N(+\infty)$$

$$\updownarrow$$

$$\forall h \exists a(h) : x > a(h) \implies f(x) > h$$

and we can say:

for every h (big as you want) there exists an a depending on h such that if $x > a(h)$ then the image of $f(x)$ is bigger than h.

As an example consider:

$$\lim_{x \to +\infty} e^x = +\infty$$

so we can write:

$$\forall h \exists a(h) : x > a(h) \implies e^x > h$$

In order to verify the correctness of the limit we should solve the inequality $e^x > h$ and obtain the result $x > a(h)$.

Now we have

$$e^x > h$$

$$\updownarrow$$

$$x > \log h = a(h)$$

Last inequality is exactly the result we were looking for.

2. $\lim_{x \to +\infty} f(x) = -\infty$

$$\forall N(-\infty) \exists J(+\infty) : x \in J(+\infty) \implies f(x) \in N(-\infty)$$

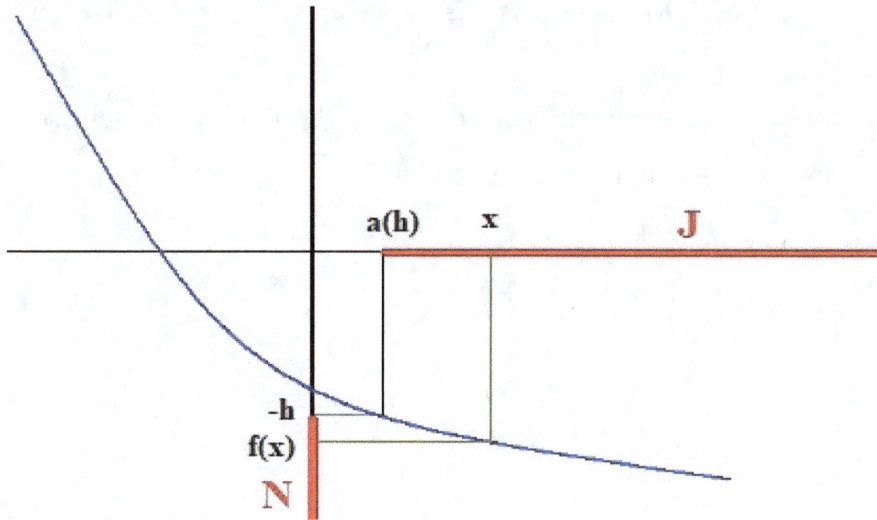

Once more please note that: $N(-\infty) =]-\infty, -h[$ (again h stands for huge) and $J(+\infty) =]a(h), +\infty[$; we write $a(h)$ because the lower bound of J is function of h. Moreover to write $x \in J(+\infty)$ is the same as to write $x > a(h)$ and to write $f(x) \in N(-\infty)$ is equivalent to write $f(x) < -h$. So we have:

$$\forall N(-\infty) \exists J(+\infty) : x \in J(+\infty) \implies f(x) \in N(-\infty)$$

$$\Updownarrow$$

$$\forall h \exists a(h) : x > a(h) \implies f(x) < -h$$

and we can say:

for every h (big as we want) there exists an a depending on h such that if $x > a(h)$ then the image of $f(x)$ is smaller than $-h$.

As an example let's consider:

$$\lim_{x \to +\infty} \log_{\frac{1}{e}} x = -\infty$$

101

so we can write:

$$\forall h \exists a(h) : x > a(h) \implies \log_{\frac{1}{e}} x < -h$$

In order to verify the correctness of the limit we should solve the inequality $\log_{\frac{1}{e}} x < -h$ and obtain the result $x > a(h)$.

Now we have

$$\log_{\frac{1}{e}} x < -h$$
$$\updownarrow$$
$$x > \left(\frac{1}{e}\right)^{-h} = e^h = a(h)$$

Last inequality is exactly the result we were looking for.

3. $\lim\limits_{x \longrightarrow +\infty} f(x) = l \quad (l \in \mathbb{R}, \text{ that is } l \text{ is not } \pm\infty)$

$$\forall N(l) \exists J(+\infty) : x \in J(+\infty) \implies f(x) \in N(l)$$

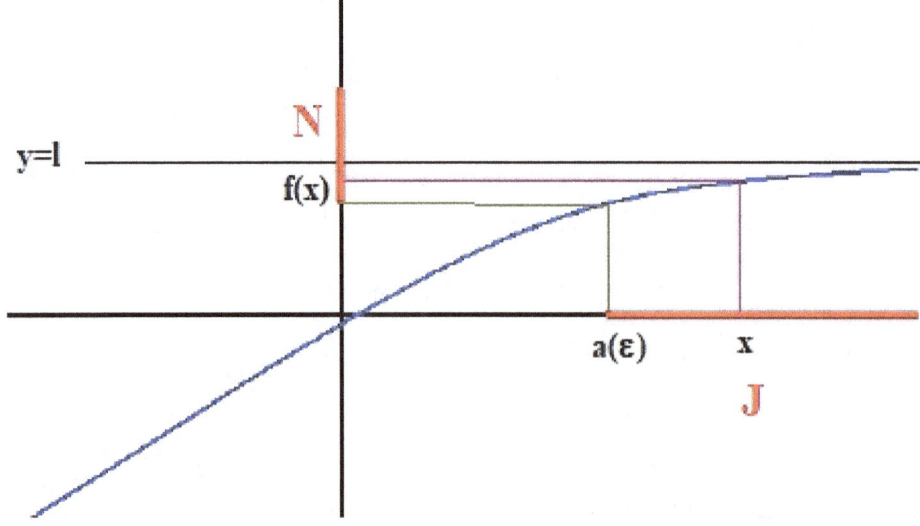

Now we can image $N(l)$ as an ϵ neighborhood centered in $l =: N(l)\,]l-\epsilon, l+\epsilon[$ (now ϵ is a positive number very close to 0) and the neighborhood J depends on ϵ: $J(+\infty) = \,]a(\epsilon), +\infty[$. Moreover to write $x \in J(+\infty)$ is the same as to write $x > a(\epsilon)$ and to write $f(x) \in N(l)$ is equivalent to write $l - \epsilon < f(x) < l + \epsilon$. Last inequalities can be written as $|f(x) - l| < \epsilon$. So we have:

$$\forall N(l)\, \exists J(+\infty) : x \in J(+\infty) \implies f(x) \in N(l)$$

$$\Updownarrow$$

$$\forall \epsilon\, \exists a(\epsilon) : x > a(\epsilon) \implies l - \epsilon < f(x) < l + \epsilon$$

and now, recalling that the absolute value can be used as a distance, we can say:

for every ϵ (positive but close to 0 as we want) there exists an a depending on ϵ such that if $x > a(\epsilon)$ then the distance of $f(x)$ from l is smaller than ϵ.

While considering next cases complete the graphs pointing out the neigborhoods, their bounds, c, x and $f(x)$.

4. $\lim\limits_{x \to -\infty} f(x) = +\infty$

$$\forall N(+\infty)\, \exists J(-\infty) : x \in J(-\infty) \implies f(x) \in N(+\infty)$$

$$\Updownarrow$$

$$\forall h\, \exists b(h) : x < b(h) \implies f(x) > h$$

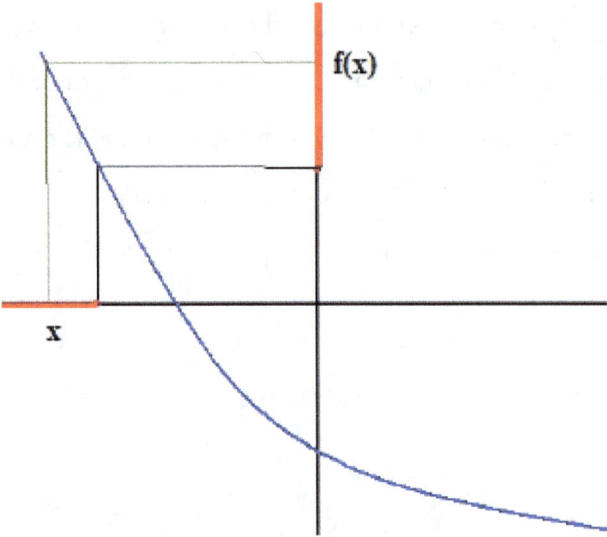

Let's consider as an example:

$$\lim_{x \longrightarrow -\infty} \left(\frac{1}{2}\right)^x = +\infty$$

$$\forall N(+\infty) \; \exists J(-\infty) : x \in J(-\infty) \implies f(x) \in N(+\infty)$$

$$\Updownarrow$$

$$\forall h \; \exists b(h) : x < b(h) \implies \left(\frac{1}{2}\right)^x > h$$

the solution of last inequality is:

$$x < \log_{\frac{1}{2}} h = b(h)$$

5. $\lim_{x \longrightarrow -\infty} f(x) = -\infty$

$$\forall N(-\infty) \; \exists J(-\infty) : x \in J(-\infty) \implies f(x) \in N(-\infty)$$

$$\Updownarrow$$

$$\forall h \; \exists b(h) : x < b(h) \implies f(x) < -h$$

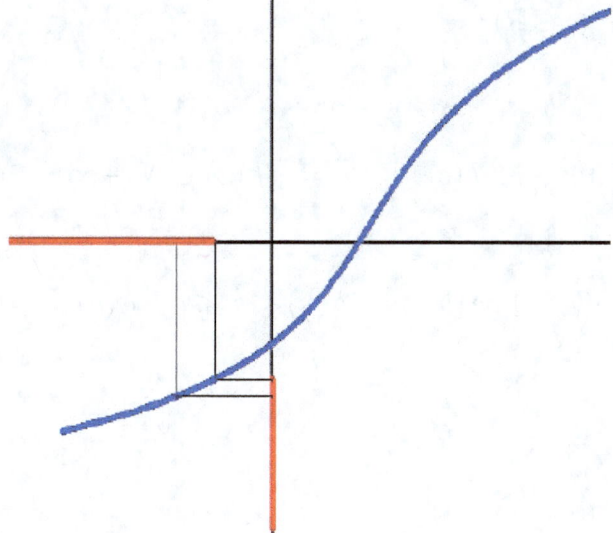

Let's consider as an example:

$$\lim_{x \longrightarrow -\infty} \sqrt[3]{x} = -\infty$$

$$\forall N(-\infty) \;\exists J(-\infty) : x \in J(-\infty) \Longrightarrow f(x) \in N(-\infty)$$

$$\Updownarrow$$

$$\forall h \;\exists b(h) : x < b(h) \Longrightarrow \sqrt[3]{x} < -h$$

the solution of last inequality is:

$$x < -h^3 = b(h)$$

6. $\lim_{x \longrightarrow -\infty} f(x) = l \qquad (l \in \mathbb{R})$

$$\forall N(l) \;\exists J(-\infty) : x \in J(-\infty) \Longrightarrow f(x) \in N(l)$$

$$\Updownarrow$$

$$\forall \epsilon > 0 \;\exists b(\epsilon) : x < b(\epsilon) \Longrightarrow l - \epsilon < f(x) < l + \epsilon$$

7. $\lim\limits_{x \longrightarrow c} f(x) = +\infty \qquad (c \in \mathbb{R})$

$$\forall N(+\infty) \; \exists J(c) : x \in J(c), x \neq c \Longrightarrow f(x) \in N(+\infty)$$

$$\Updownarrow$$

$$\forall h > 0 \; \exists \delta(h) : c - \delta(h) < x < c + \delta(h), x \neq c \Longrightarrow f(x) > h$$

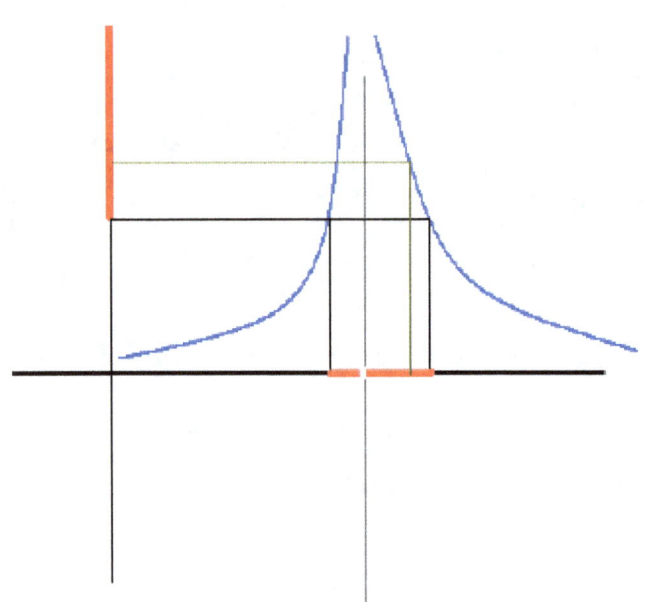

8. $\lim\limits_{x \longrightarrow c} f(x) = -\infty \qquad (c \in \mathbb{R})$

$$\forall N(-\infty) \; \exists J(c) : x \in J(c), x \neq c \Longrightarrow f(x) \in N(-\infty)$$

$$\Updownarrow$$

$$\forall h > 0 \; \exists \delta(h) : c - \delta(h) < x < c + \delta(h), x \neq c \Longrightarrow f(x) < -h$$

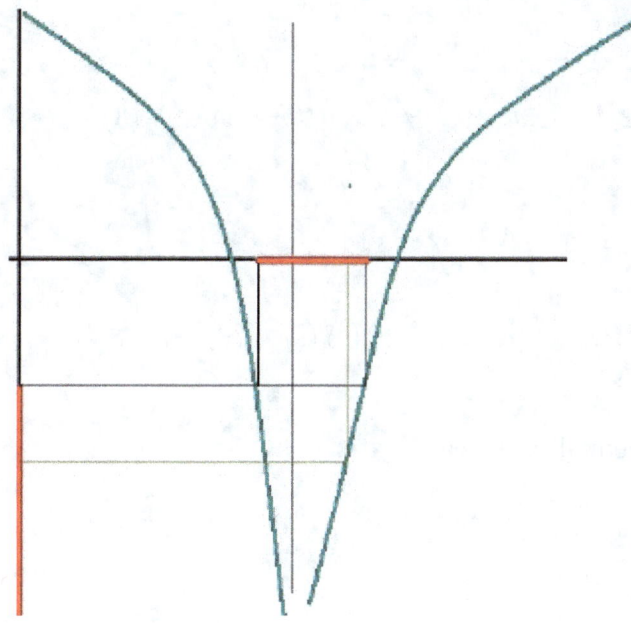

9. $\lim\limits_{x \longrightarrow c} f(x) = l \qquad (c, l \in \mathbb{R})$

$$\forall N(l) \ \exists J(c) : x \in J(c), x \neq c \Longrightarrow f(x) \in N(l)$$

$$\Updownarrow$$

$$\forall \epsilon > 0 \ \exists \delta(\epsilon) : c - \delta(\epsilon) < x < c + \delta(\epsilon), x \neq c \Longrightarrow l - \epsilon < f(x) < l + \epsilon$$

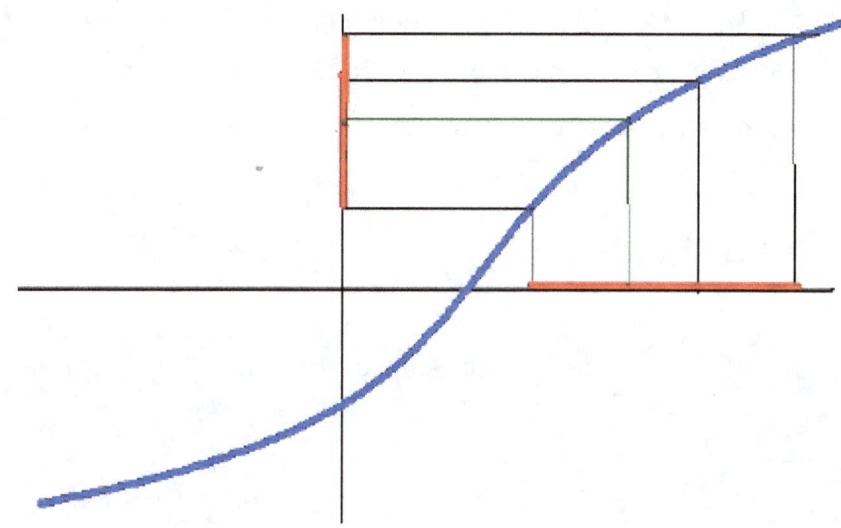

$$\lim_{x \to 2} (x+1) = 3$$

$$\forall N(3) \; \exists J(2) : x \in J(2), x \neq 2 \implies (x+1) \in N(3)$$

$$\Updownarrow$$

$$\forall \epsilon > 0 \; \exists \delta(\epsilon) : 3 - \delta(\epsilon) < x < 3 + \delta(\epsilon), x \neq 2 \implies 3 - \epsilon < x + 1 < 3 + \epsilon$$

considering last inequality we can write:

$$2 - \epsilon < x < 2 + \epsilon$$

$$\Updownarrow$$

$$3 - \delta(\epsilon) < x < 3 + \delta(\epsilon)$$

In cases 7, 8 and 9 we have considered the case $x \longrightarrow c$ meaning that x is approaching c from both sides (from the right and the left). In the following we are considering the cases when x approaches c from one side only. Now we are writing $x \longrightarrow c^+$ to say that x approaches c from the right (the simbol "+" as superscript means that we are considering values greater than c); we are writing $x \longrightarrow c^-$ to say that x approaches c from the left (the simbol "−" as superscript means that we are considering values smaller than c).

10. $\lim_{x \to c^+} f(x) = +\infty \qquad (c \in \mathbb{R})$

$$\forall N(+\infty) \; \exists J^+(c) : x \in J^+(c) \implies f(x) \in N(+\infty)$$

$$\Updownarrow$$

$$\forall h > 0 \; \exists \delta(h) : c < x < c + \delta(h) \implies f(x) > h$$

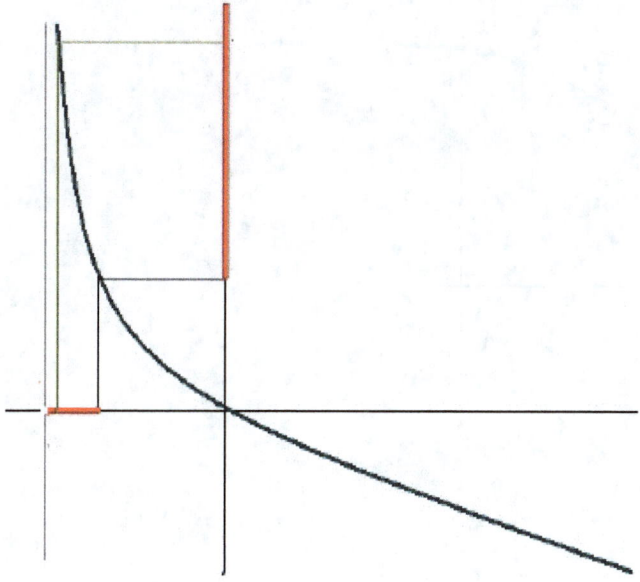

11. $\lim\limits_{x \longrightarrow c^+} f(x) = -\infty \qquad (c \in \mathbb{R})$

$$\forall N(-\infty) \; \exists J^+(c) : x \in J^+(c) \implies f(x) \in N(-\infty)$$

$$\Updownarrow$$

$$\forall h > 0 \; \exists \delta(h) : c < x < c + \delta(h) \implies f(x) < -h$$

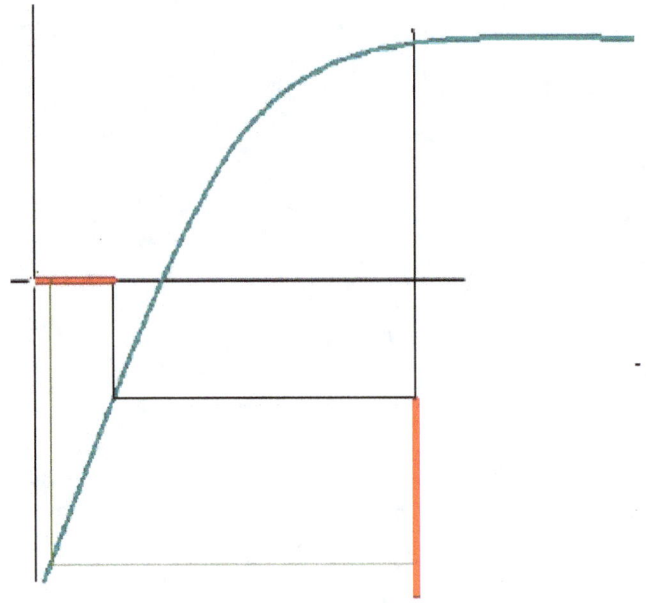

12. $\lim\limits_{x \longrightarrow c^+} f(x) = l \qquad (c, l \in \mathbb{R})$

$$\forall N(l) \ \exists J^+(c) : x \in J^+(c) \Longrightarrow f(x) \in N(l)$$

$$\Updownarrow$$

$$\forall \epsilon > 0 \ \exists \delta(\epsilon) : c < x < c + \delta(\epsilon) \Longrightarrow l - \epsilon < f(x) < l + \epsilon$$

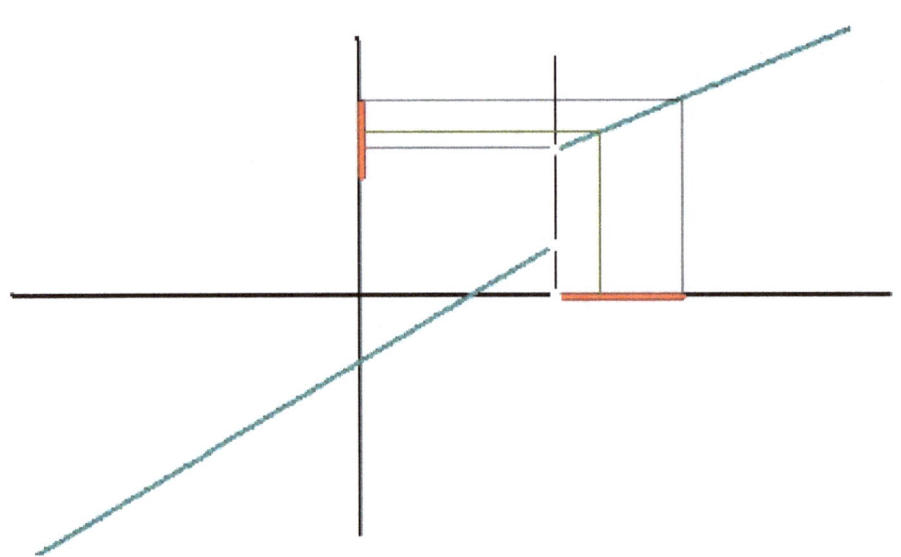

13. $\lim_{x \longrightarrow c^-} f(x) = +\infty \qquad (c \in \mathbb{R})$

$$\forall N(+\infty) \; \exists J^-(c) : x \in J^-(c) \Longrightarrow f(x) \in N(+\infty)$$

$$\Updownarrow$$

$$\forall h > 0 \; \exists \delta(h) : c - \delta(h) < x < c \Longrightarrow f(x) > h$$

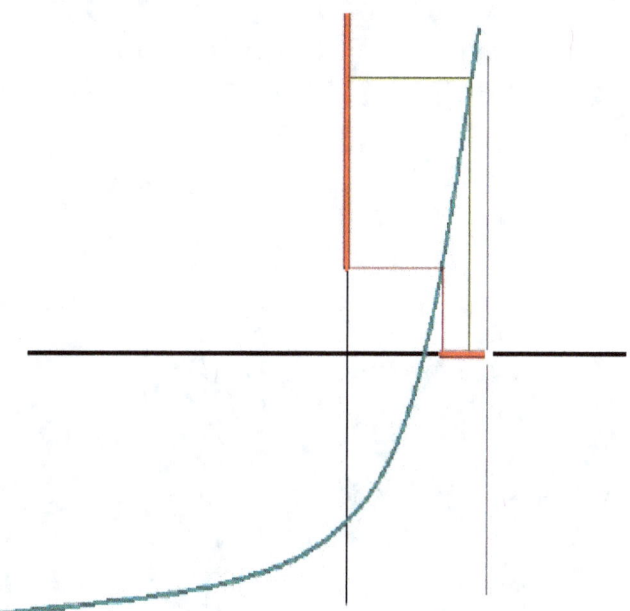

14. $\lim_{x \longrightarrow c^-} f(x) = -\infty \qquad (c \in \mathbb{R})$

$$\forall N(-\infty) \; \exists J^-(c) : x \in J^-(c) \Longrightarrow f(x) \in N(-\infty)$$

$$\Updownarrow$$

$$\forall h > 0 \; \exists \delta(h) : c - \delta(h) < x < c \Longrightarrow f(x) < -h$$

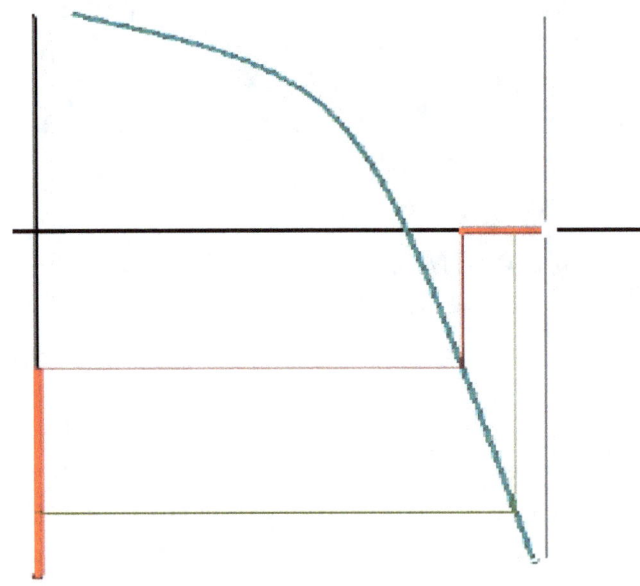

15. $\lim_{x \longrightarrow c^-} f(x) = l \qquad (c, l \in \mathbb{R})$

$$\forall N(l) \ \exists J^-(c) : x \in J^-(c) \Longrightarrow f(x) \in N(l)$$

$$\Updownarrow$$

$$\forall \epsilon > 0 \ \exists \delta(\epsilon) : c - \delta(\epsilon) < x < c \Longrightarrow l - \epsilon < f(x) < l + \epsilon$$

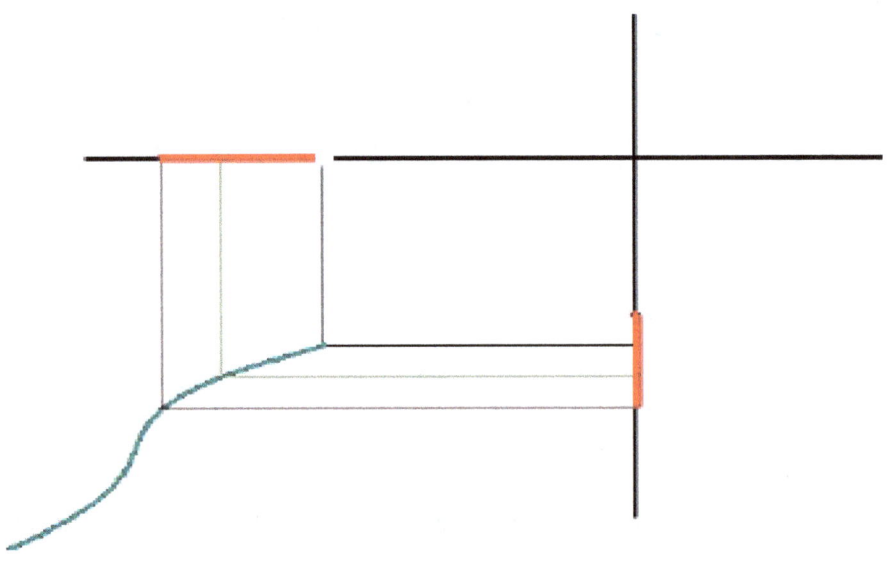

If a function satisfies the definition in case 3 or 6 then the line $y = l$ is called horizontal asymptote.

If a function satisfies one among the definitions 7, 8, 10, 11, 13, 14 then the line $x = c$ is called vertical asymptote.

Chapter 8

Theorems on limits

In this chapter I introduce the main theorems concerning the limits; they will be useful for the limit computation and for next topics will face in this book.

Theorem 8.1 *Uniqueness of limits: if a limit exists it is unique:*

$$\left.\begin{array}{l}\lim_{x \to c} f(x) = l_1 \\ \lim_{x \to c} f(x) = l_2\end{array}\right\} \Longrightarrow l_1 = l_2$$

Proof. We prove the theorem by contradiction; first let's deny the consequence:

$$l_1 \neq l_2 \Longrightarrow \begin{cases} \forall N(l_1) \, \exists J_1(c) : x \in J_1(c), \, x \neq c, \Longrightarrow f(x) \in N(l_1) \\ \forall N(l_2) \, \exists J_2(c) : x \in J_2(c), \, x \neq c, \Longrightarrow f(x) \in N(l_2) \end{cases}$$

we can choose any $N(l_1)$ and $N(l_2)$ and they can be disjoint that is $N(l_1) \cap N(l_{1_2}) = \varnothing$; the same cannot happen for $J_1(c)$ and $J_2(c)$; hence if:

$$x \in J_1(c) \cap J_2(c) \Longrightarrow \begin{cases} f(x) \in N(l_1) \\ f(x) \in N(l_2) \end{cases}$$

but this is not possible because a function cannot assign two different images to an element of the domain. This contradiction proves the theorem. ∎

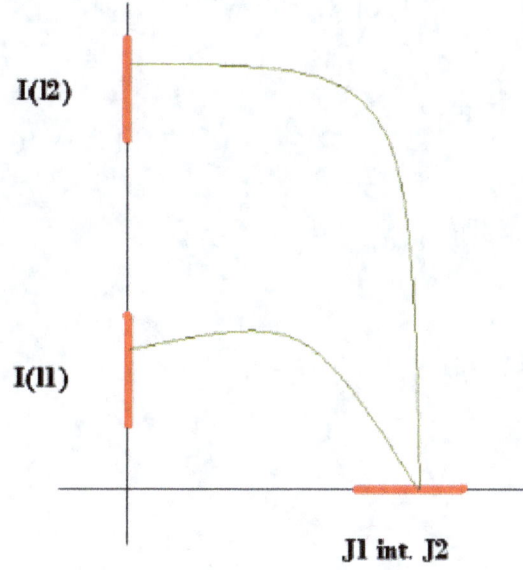

Theorem 8.2 *Comparison theorem: let f(x) and g(x) be two functions existing in a neighborhood $N(c) - \{c\}$ and $\forall x \in N(c) - \{c\}$ we have $f(x) \leq g(x)$; then:*

$$\lim_{x \to c} f(x) = +\infty \implies \lim_{x \to c} g(x) = +\infty$$

$$\lim_{x \to c} g(x) = -\infty \implies \lim_{x \to c} f(x) = -\infty$$

Proof. (first part)

$$\lim_{x \to c} f(x) = +\infty \iff \forall N(+\infty) \, \exists J(c) : x \in J(c), x \neq c \implies f(x) \in N(+\infty)$$

Since $\forall N(+\infty) =]a, +\infty[$ we can write:

$$\lim_{x \to c} f(x) = +\infty$$

$$\Updownarrow$$

$$\forall a \exists x(a) : x \geq x(a) \implies f(x) > a$$

$$\Downarrow$$

$$g(x) \geq f(x) > a$$

$$\Updownarrow$$

$$\lim_{x \to c} g(x) = +\infty$$

∎

The following graph shows a particular case of the theorem:

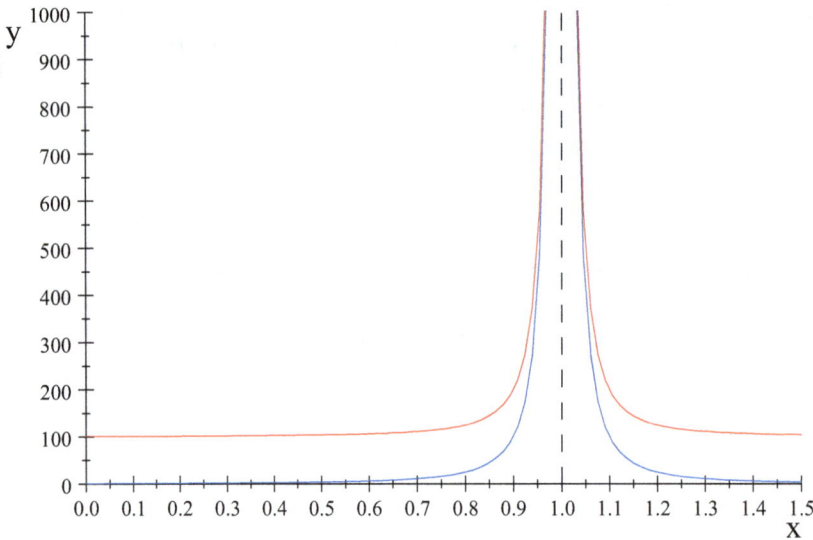

The functions in the graph are: $f(x) = \frac{1}{(x-1)^2}$ (in blue) and $g(x) = \frac{1}{(x-1)^2} + 100$ (in red).

Theorem 8.3 *Squeeze theorem (also known as "sandwich theorem", "two policemen and a drunk theorem", "teorema dei due carabinieri" in Italian): Let f(x), g(x) and h(x) be three functions exisisting in $N(c) - \{c\}$ and such that $\forall x \in N(c) - \{c\}$ we have $f(x) \leq g(x) \leq h(x)$; then:*

$$\left[\lim_{x \to c} f(x) = \lim_{x \to c} h(x) = l\right] \implies \left[\lim_{x \to c} g(x) = l\right]$$

Proof.

- if $l = +\infty$ then using the comparison theorem we can conclude that $\lim_{x \to c} g(x) = +\infty$

- if $l = -\infty$ then using the comparison theorem we can conclude that $\lim_{x \to c} g(x) = -\infty$

- if $l \in \mathbb{R}$ then:

$$\left[\lim_{x \to c} f(x) = l\right] \iff [\forall N(l) \, \exists J_f(c) : x \in J_f(c), x \neq c \implies f(x) \in N(l)]$$

$$\left[\lim_{x \to c} h(x) = l\right] \iff [\forall N(l) \, \exists J_h(c) : x \in J_h(c), x \neq c \implies h(x) \in N(l)]$$

a neighborhood of l (finite number) is an interval like $]l - \epsilon, l + \epsilon[$ and we can write the above definitions as follows:

$$\left[\lim_{x \to c} f(x) = l\right] \iff [\forall \epsilon > 0 \, \exists J_f(c) : x \in J_f(c), x \neq c \implies l - \epsilon < f(x) < l + \epsilon]$$

$$\left[\lim_{x \to c} h(x) = l\right] \iff [\forall \epsilon > 0 \, \exists J_h(c) : x \in J_h(c), x \neq c \implies l - \epsilon < h(x) < l + \epsilon]$$

then $\forall \epsilon > 0$:

$$x \in J_f(c) \cap J_h(c)$$
$$\Downarrow$$
$$l - \epsilon < f(x) \leq g(x) \leq h(x) < l + \epsilon$$

and last formula is the definition of $\lim_{x \to c} g(x) = l$

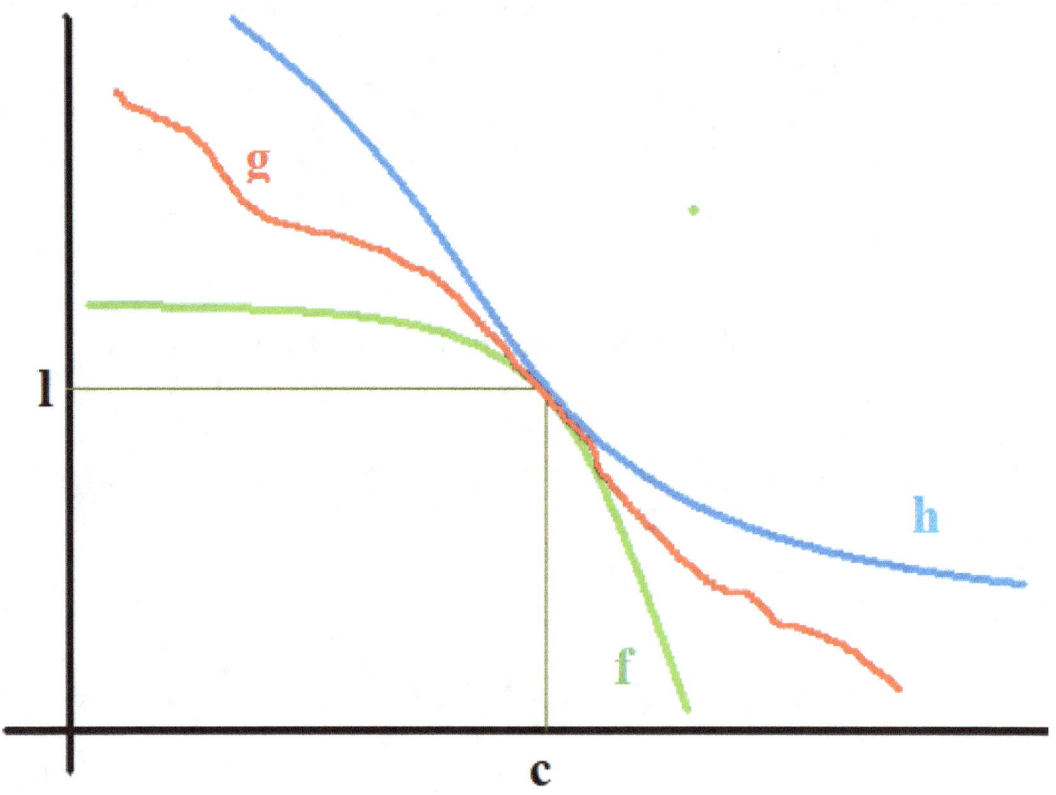

Graphical representation of the squeeze theorem

Example 8.1 1. $\lim\limits_{x\to+\infty} \cos x$ *does not exist but:*

$$\lim_{x\to+\infty} \frac{\cos x}{x} = 0$$

since $-1 \leq \cos x \leq +1$ and (if $x > 0$)

$\frac{-1}{x} \leq \frac{\cos x}{x} \leq \frac{+1}{x}$ and

$\lim\limits_{x\to+\infty} \frac{-1}{x} = \lim\limits_{x\to+\infty} \frac{+1}{x} = 0.$

2. $\lim\limits_{x\to 0} \sin \frac{1}{x}$ does not exist; the following is the (very strange) graph of the function $f(x) = \sin \frac{1}{x}$

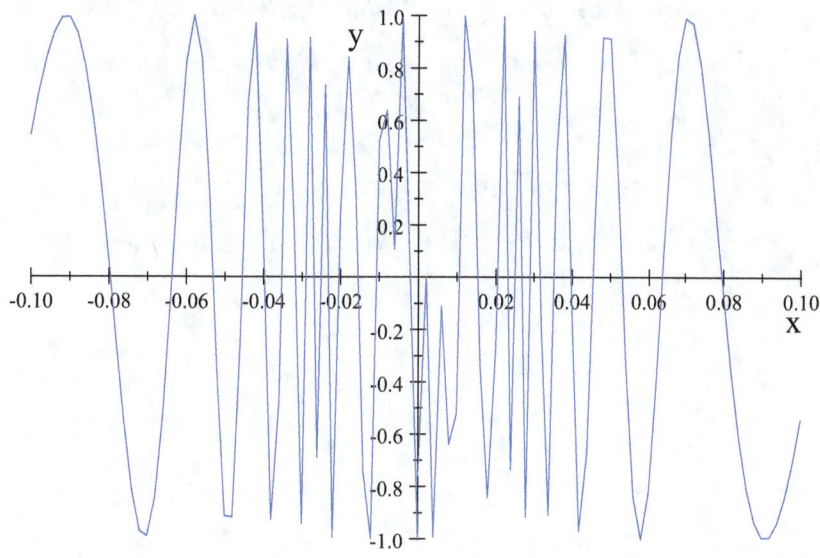

the function oscillates infinite times between -1 and +1 in each neighborhood of 0 (the graph is not so helpful in this case).

But, because of the squeeze theorem, we can conclude that:

$$\lim_{x \to 0} x \sin \frac{1}{x} = 0$$

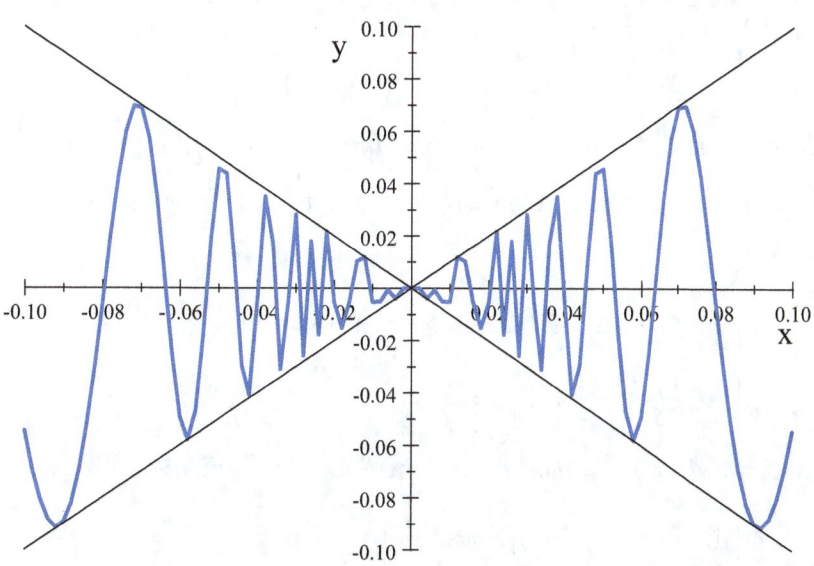

The two lines in the picture have equation $y = x$ and $y = -x$ and we can see that the function $f(x) = x \sin \frac{1}{x}$ is bounded between them and tends to 0 as x approaches 0.

By a similar argument we can show that $\lim\limits_{x \to 0} x^2 \sin \frac{1}{x} = 0$ (see picture below and realize how the "wild" function $\sin \frac{1}{x}$ is squeezed between the two parabolas $y = x^2$ and $y = -x^2$).

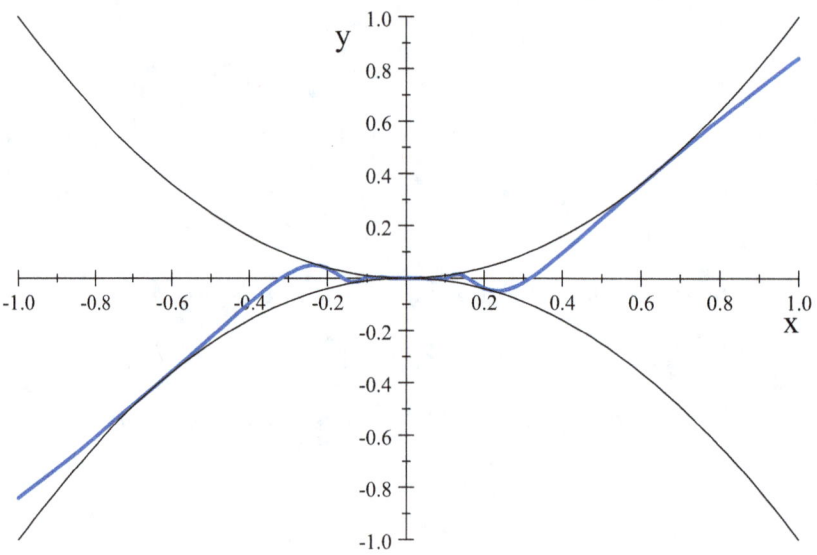

the parabolas' equations are: $y = x^2$ and $y = -x^2$ and we have $-x^2 \leq x^2 \sin \frac{1}{x} \leq x^2$.

Theorem 8.4 *Sign Theorem: if the limit of a function exists in c and it is not zero then there exists a neighborhood of c where the function has the same sign of the limit:*

$$\left[\lim_{x \to c} f(x) = l > 0\right] \implies [\exists J(c) : x \in J(c), x \neq c \implies f(x) > 0]$$

$$\left[\lim_{x \to c} f(x) = l < 0\right] \implies [\exists J(c) : x \in J(c), x \neq c \implies f(x) < 0]$$

We can state the theorem also in another (equivalent) way:

$$\left[\lim_{x \to c} f(x) = l \neq 0\right] \implies [\exists J(c) : x \in J(c), x \neq c \implies f(x) \cdot l > 0]$$

Proof. Let's prove the theorem when $l > 0$, the proof in the case $l < 0$ is similar.

$$\lim_{x \to c} f(x) = l$$

$$\Updownarrow$$

$$\forall N(l) \, \exists J(c) : x \in J(c), x \neq c \Longrightarrow f(x) \in N(l)$$

We can choose $N(l)$ as we want, so we can choose a neighborhood fully contained in $\mathbb{R}^+ - \{0\}$; in such a case

$$\forall x \in J(c), x \neq c \Longrightarrow f(x) \in N(l) \subset \mathbb{R}^+ - \{0\}$$

and this means that $f(x) > 0$.

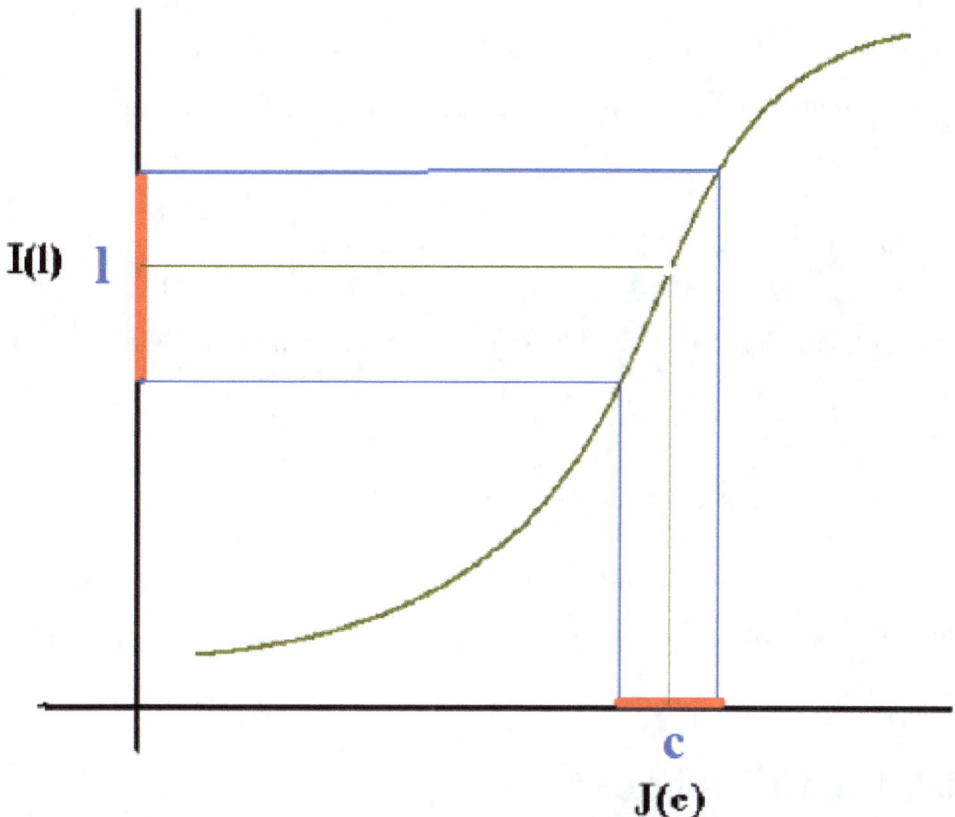

■

Chapter 9

Limits computation

To compute $\lim\limits_{x \to c} f(x)$ we have first to substitute c in the place of x in the function and (try to) compute the value following the rules of elementary algebra; as an example we can compute

$$\lim_{x \to 2} \frac{1}{1-x} = -1$$

But the rules of the traditional algebra are not enough! think as an example to

$$\lim_{x \to 1} \frac{1}{1-x} = \frac{1}{0}$$

and of course it is not possible to compute $\frac{1}{0}$.

For this reason we have to introduce "new" computation rules and some new techniques to achieve our goal.

9.1 Partial Algebra

In this section we try to apply the rules of the traditional algebra to infinite (very big numbers) and infinitesimal (numbers very close to zero); unfortunately it is not possible to apply the operators to all the cases we may face. Thats why we don't have an algebra but only a "partial algebra".

First of all we need to introduce a formal definition for infinite and infinitesimal.

Definition 9.1 *Infinite and infinitesimal*

Let f be a function $\mathbb{R} \longrightarrow \mathbb{R}$ we say that $f(x)$ is an infinite when $x \longrightarrow c$ if

$$\lim_{x \to c} f(x) = \infty$$

we say that $f(x)$ is an infinitesimal when $x \longrightarrow c$ if

$$\lim_{x \to c} f(x) = 0$$

We use the following notation:

- ∞ to denote an infinite (we don't specify the sign);

- $\pm\infty$ to denote an infinite (positive or negative);

- 0 to denote an infinitesimal different by 0 (positive or negative);

- 0^+ to denote a positive infinitesimal different by 0;

- 0^- to denote a positive infinitesimal different by 0;

- k is a positive number different by 0 that is neither an infinitesimal nor an infinite.

Now I am going to present the main rules of this partial algebra; the complete proofs are not presented but I strongly recommend the reader to verify the rules using numerical examples; in some of the cases I will help the reader with some examples.

Keep in mind that all the rules of the traditional algebra still holds; in particular the wellknown "sign rule" is still valid ("+ · − = −" and so on).

Considering the elementary operators "+" and "·" we can introduce the following computation rules:

$$+\infty \pm k = +\infty$$
$$-\infty \pm k = -\infty \tag{9.1}$$

The idea underlying this rule is quite simple: imagine adding (or subtracting) 1 to infinity, it is the same thing as to pour (or detract) a glass of water in the ocean! Nothing changes, this is why we can conclude that $+\infty+1 = +\infty$ and $+\infty-1 = +\infty$.

$$k \cdot \infty = \infty \tag{9.2}$$

$$\frac{k}{\infty} = 0 \tag{9.3}$$

About last rule try (using a calculator if needed) to compute $\frac{1}{10}$, $\frac{1}{100}$, $\frac{1}{1000}$ and so on; you should immediately realize that when the denominator becomes bigger and bigger the fraction becomes closer and closer to zero; since we can imagine to add infinity many zeros to the denominator we can get closer and closer to zero. Moreover, using an even simpler argument: let's image to share a cake (I like cakes very much) among all the people leaving in the European community! Each guy will receive almost nothing and nothing in mathematics is 0.

$$\frac{k}{0^{\pm}} = \pm\infty \tag{9.4}$$

Once more try to compute $\frac{1}{0.1}$, $\frac{1}{0.01}$, $\frac{1}{0.001}$ and so on; as you can see when the denominator gets closer to 0 the fraction increases.

$$\frac{\infty}{0} = \infty \tag{9.5}$$

$$\frac{0}{\infty} = 0 \tag{9.6}$$

Rule (9.5) is a consequence of rules (9.2) and (9.4); rule (9.6) is a consequence of fraction properties ($\frac{0}{k} = 0 \forall k$) and rule (9.3).

Example 9.1 *(concerning relation (9.3))*

Let's verify

$$\lim_{x \to +\infty} \frac{1}{x} = 0$$

$$\Updownarrow$$

$$\forall N(+\infty) \, \exists J(+\infty) : x \in J(+\infty) \implies \frac{1}{x} \in N(0)$$

Since we can write $N(0) =]-\epsilon, +\epsilon[$ and $J(+\infty) =]h, +\infty[$ the above limit definition is equivalent to:

$$\forall \epsilon > 0 \, \exists h(\epsilon) : x > h(\epsilon) \implies -\epsilon < \frac{1}{x} < \epsilon$$

In order to verify the limit definition we have to solve the inequalities $-\epsilon < \frac{1}{x} < \epsilon$ which can be written as

$$\begin{cases} \frac{1}{x} < \epsilon \\ \frac{1}{x} > -\epsilon \end{cases}$$

Since we are considering $x \to +\infty$ we can assume x positive and so the second inequality is certainly satisfied ($\frac{1}{x}$ is certainly greater than a negative number);

$$\frac{1}{x} < \epsilon \quad \iff \quad x > \frac{1}{\epsilon}$$

and the checking of the definition is complete since to write $-\frac{1}{\epsilon} < x < \frac{1}{\epsilon}$ is the same as to write $x \in N_\epsilon(0)$.

Example 9.2 *(concerning relation (9.4))*

Let's verify

$$\lim_{x \to 0^+} \frac{1}{x} = +\infty$$

$$\Updownarrow$$

$$\forall N(+\infty) \, \exists J^+(0) : x \in J^+(0) \implies \frac{1}{x} \in N(+\infty)$$

Since we can write $N(+\infty) =]a, +\infty[$ and $J^+(0) =]0, \epsilon[$ the above limit definition is equivalent to:

$$\forall a \exists x(a) : 0 < x < x(a) \implies \frac{1}{x} > a$$

In order to verify the limit definition we have to solve the inequality $\frac{1}{x} > a$ by considering only positive value for variable x. The solution is

$$x < \frac{1}{a}$$

and the checking of the definition is complete since to write $0 < x < \frac{1}{a}$ is the same as to write $x \in J^+(0)$.

Please note that we have

$$\lim_{x \to 0^-} \frac{1}{x} = -\infty$$

and that we can also compute

$$\lim_{x \to 0} \frac{1}{x} = \infty$$

In last example, since we are not specifying if x is approaching 0 from the left or from the right we cannot determine if the result is $+\infty$ or $-\infty$. Next picture shows the graph of function $y = \frac{1}{x}$

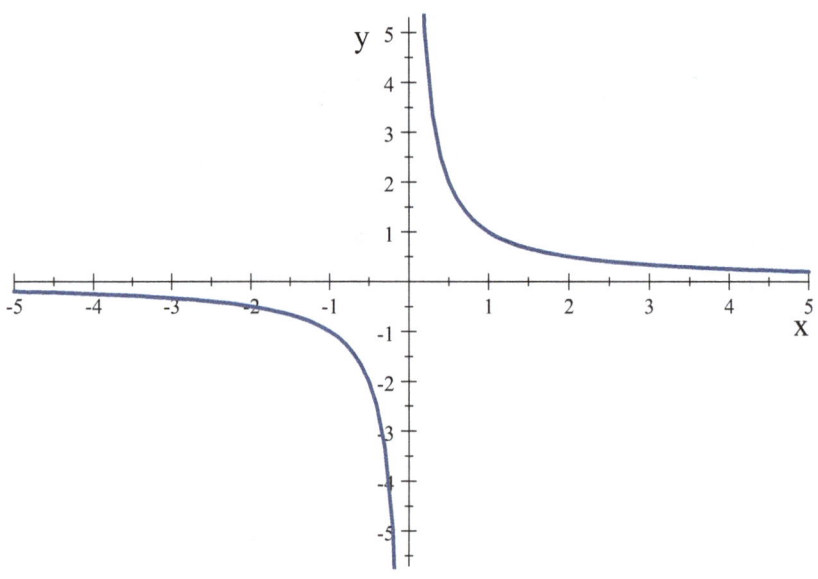

Here it is possible to note that if x approaches 0 from the right the function becomes bigger and bigger while approaching the same point from the left it becomes more an more negative. From the graph we can also note that when x tends to positive or negative infinity the function value becames closer and closer to 0.

Keeping in mind power properties we can sense the following rules:

$$\left.\begin{array}{rcl} k^{+\infty} &=& +\infty \\ k^{-\infty} &=& 0 \end{array}\right\} \text{ if } k > 1 \qquad (9.7)$$

and once more I suggest to the reader to check numerically these rules (if necessary). About the second rule consider, as an example 10^x and assign to the exponent the values $-1, -2, -3....$ Moreover recalling that $a^{-x} = \frac{1}{a^x}$ we have

$$10^{-\infty} = \frac{1}{10^{+\infty}} = 0$$

$$\left.\begin{array}{rcl} k^{+\infty} &=& 0 \\ k^{-\infty} &=& +\infty \end{array}\right\} \text{ if } 0 < k < 1 \qquad (9.8)$$

(about last relation if necessary consider numerical examples with $k = 0.9$).

$$\left(0^+\right)^{+\infty} = 0 \qquad (9.9)$$

I bet you are not surprised.

$$\left(0^+\right)^{-\infty} = +\infty \qquad (9.10)$$

I bet you are really surprised! But try to compute 0.001^{-1}, 0.001^{-2} and so on.

Moreover

$$\left.\begin{array}{rcl} (+\infty)^k &=& +\infty \\ (+\infty)^{-k} &=& 0 \end{array}\right. \qquad (9.11)$$

At the end of this section I ask the reader to discover the following rules

$$(+\infty)^{+\infty} = ?$$
$$(+\infty)^{-\infty} = ?$$

Exercise 9.1 *verify the following:*

$$\lim_{x \to +\infty} e^x = +\infty$$

$$\lim_{x \to -\infty} e^x = 0$$

9.2 Missing (Unlucky) Cases

Formally speaking the following are called "indeterminate cases", in my mind the best way to denote this cases is "missing cases" since we don't have any rules to immediately determine the result. Because of this cases we cannot complete the rules underlying the operators and this is why we have (only) a partial algebra. We can distinguish two categories of missing cases: the ones involving the operators "+" and "·" (which we will refer to as algebraic cases) and the ones involving exponents (exponential cases).

9.2.1 Algebraic cases

1.

$$\boxed{(+\infty - \infty) \xrightarrow{?} \begin{cases} +\infty \\ l \in \mathbb{R} \\ -\infty \end{cases}}$$

When computing the difference $+\infty - \infty$ we might obtain one of the result listed above; let's consider three different examples:

- $\lim\limits_{x \to +\infty^-} (x^3 - x^2) = +\infty - \infty$

  It is possible to solve this case using simple manipulations:

  $$\lim\limits_{x \to +\infty^-} x^2(x-1) = (+\infty)(+\infty - 1) = +\infty$$

  We got this result since when $x \to +\infty$ x^3 is much (much much) bigger than x^2.

- $\lim\limits_{x \to +\infty^-} (x^3 - x^4) = +\infty - \infty = x^3(1-x) = (+\infty)(1-\infty) = -\infty$ (this happens since x^4 is much bigger than x^3)

- $\lim\limits_{x \to +\infty^-} \left(\sqrt{x+1} - \sqrt{x}\right) = +\infty - \infty = \lim\limits_{x \to +\infty^-} \left(\sqrt{x+1} - \sqrt{x}\right) \cdot \frac{\sqrt{x+1}+\sqrt{x}}{\sqrt{x+1}+\sqrt{x}} =$
  $\lim\limits_{x \to +\infty^-} \frac{\left(\sqrt{x+1}\right)^2 - \left(\sqrt{x}\right)^2}{\sqrt{x+1}+\sqrt{x}} = \lim\limits_{x \to +\infty^-} \frac{x+1-x}{\sqrt{x+1}+\sqrt{x}} = \frac{1}{+\infty+\infty} = 0$

  The result is 0 since when x is a really big number $\sqrt{x+1}$ is really close to $\sqrt{x}$.

In these examples we succesfully solved the problems combining the rules of the traditional algebra with the rules of the partial algebra; we'll see that sometimes these rules are not enough.

2.

$$\frac{\infty}{\infty} \xrightarrow{?} \begin{cases} \infty \\ l \in \mathbb{R}, l \neq 0 \\ 0 \end{cases}$$

Let's consider three more examples:

- $\lim\limits_{x \to +\infty} \frac{x^3}{-x^2} = \frac{+\infty}{-\infty} = \lim\limits_{x \to +\infty^-} \frac{x}{-1} = \frac{+\infty}{-1} = -\infty$

- $\lim\limits_{x \to +\infty} \frac{x^3}{-x^4} = \frac{+\infty}{-\infty} = \lim\limits_{x \to +\infty^-} \frac{1}{-x} = \frac{1}{-\infty} = 0$

- $\lim\limits_{x \to +\infty} \frac{x+1}{x} = \frac{+\infty}{-\infty} = \lim\limits_{x \to +\infty^-} \left(1 + \frac{1}{x}\right) = 1 + 0 = 1$

3.

$$\boxed{0 \cdot \infty \xrightarrow{?} \begin{cases} \infty \\ l \in \mathbb{R}, l \neq 0 \\ 0 \end{cases}}$$

This is another missing case since the result may be ∞ (from the partial algebra we know that $k \cdot \infty = \infty$) or 0 (from the traditional algebra $0 \cdot x = 0 \forall x$) or something else.

Remark 9.1 *in the indeterminate form $0 \cdot \infty$ the first factor is NOT 0 but a number very close to 0. If we are computing*

$$\lim_{x \to +\infty} e^{-x} \cdot x$$

we are dealing with the missing case $0 \cdot \infty$ since e^{-x} is a quantity very close to 0 when $x \to +\infty$. If we are considering

$$\lim_{x \to +\infty} 0x$$

we are not considering an indeterminate form since the first factor is 0 (not something that tends to 0). So in this case we have:

$$\lim_{x \to +\infty} 0x = 0$$

4.

$$\boxed{\frac{0}{0} \xrightarrow{?} \begin{cases} 0 \\ l \in \mathbb{R}, l \neq 0 \\ \infty \end{cases}}$$

As in the above case the simble 0 denotes a non zero quantity very close to 0.

- $\lim_{x \to 0^-} \frac{x^3}{-x^2} = \frac{0}{0} = \lim_{x \to 0^-} \frac{x}{-1} = \frac{0}{-1} = 0$

- $\lim_{x\to 0^+} \frac{x^3}{-x^4} = \frac{0}{0} = \lim_{x\to 0^+} \frac{1}{-x} = \frac{1}{0^-} = -\infty$

- $\lim_{x\to -\infty} \frac{2^{x+1}}{2^x} = \frac{0}{0} = \lim_{x\to +\infty} \frac{2^x \cdot 2}{2^x} = 2$

Let me remind you that $\frac{\infty}{0}$ and $\frac{0}{\infty}$ are not are not indeterminate cases.

9.2.2 Exponential cases

1.

$$(+\infty)^0 \xrightarrow{?} \begin{cases} 1 \\ l \in \mathbb{R}, l \neq 1 \text{ and } l \neq 0 \\ +\infty \text{ or } 0 \end{cases}$$

The ambiguity comes from the rule of the traditional algebra $k^0 = 1$ and the rules of the partial algebra $(+\infty)^k = +\infty$ and $(+\infty)^{-k} = 0$.

2.

$$(0^+)^0 \xrightarrow{?} \begin{cases} 1 \\ l \in \mathbb{R}, l \neq 1 \text{ and } l \neq 0 \\ +\infty \text{ or } 0 \end{cases}$$

Now the ambiguity comes from the traditional algebra rules $k^0 = 1$ and $0^k = 0$.

3.

$$(1)^\infty \xrightarrow{?} \begin{cases} +\infty \text{ or } 0 \\ l \in \mathbb{R}, l \neq 1 \text{ and } l \neq 0 \\ 1 \end{cases}$$

9.3 Fundamental limits

We can solve the exponential case 3 using a fundamental limit (it is really important):

$$\boxed{\lim_{x \to \infty} \left(1 + \tfrac{1}{x}\right)^x = e \approx 2.7}$$

$$e \in \mathbb{R} \qquad e \notin \mathbb{Q}$$

e is an irrational number with many nice properties; later in this book we'll see some of these properties.

I don't present the proof of the result since it requires some tools not covered in this book. Using the fundamental limit we can obtain a large set of significant limits; using simple substitutions its easy to prove that:

$$\lim_{x \to \infty} \left(1 - \tfrac{1}{x}\right)^x = e^{-1} = \tfrac{1}{e}$$

$$\lim_{x \to 0} \left(1 + x\right)^{\tfrac{1}{x}} = e$$

$$\lim_{x \to 0} \left(1 - x\right)^{\tfrac{1}{x}} = e^{-1} = \tfrac{1}{e}$$

We can face the unlucky case $\tfrac{0}{0}$ by factorization (as we've done in the previous section) or using a set of significant limits; such limits are derived from the fundamental limit presented above and from the following fundamental limit:

$$\boxed{\lim_{x \to 0} \tfrac{\sin x}{x} = 1}$$

We are going to prove the above result considering only the case $x > 0$ (when x is negative the proof is almost the same). Consider the picture:

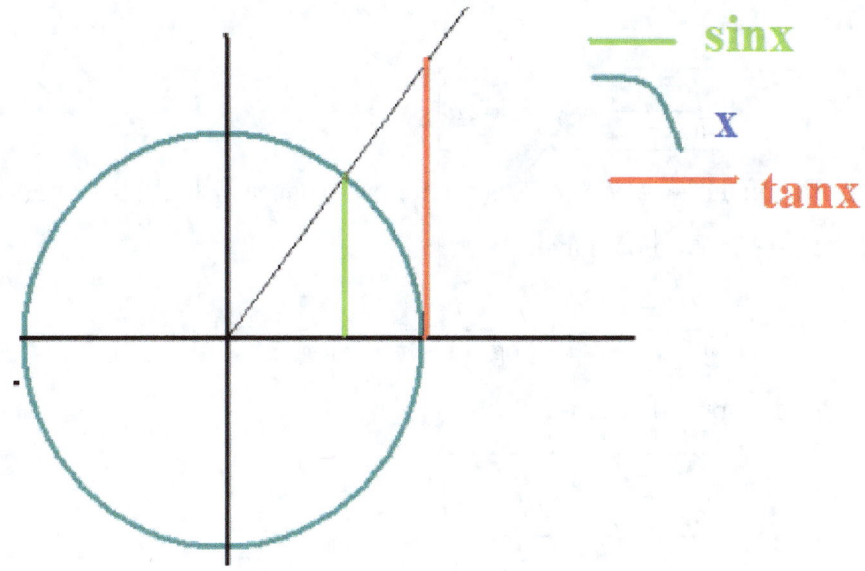

$$\sin x \leq x \leq \tan x$$
$$\frac{\sin x}{\sin x} \leq \frac{x}{\sin x} \leq \frac{\tan x}{\sin x}$$
$$1 \leq \lim_{x \to 0} \frac{x}{\sin x} \leq \lim_{x \to 0} \frac{1}{\cos x} = 1$$

We get the solution applying the squeezing theorem. From the (two) fundamental limits above we can derive many significant limits:

1.
$$\boxed{\lim_{x \to 0} \frac{\log_a(1+x)}{x} = \frac{0}{0} = \log_a e}$$

Proof: $\lim_{x \to 0} \frac{\log_a(1+x)}{x} = \frac{1}{x} \log_a(1+x) = \log_a(1+x)^{\frac{1}{x}} = \log_a e$

and a special case of this notable limit is

$$\boxed{\lim_{x \to 0} \frac{\log(1+x)}{x} = \log e = 1}$$

(let me remind you that $\log x$ is $\log_e x$).

2.
$$\boxed{\lim_{x \to 0} \frac{a^x - 1}{x} = \frac{0}{0} = \log a}$$

Proof: let's set $t = a^x - 1$; it follows that $x = \log_a (1 + t)$ and if $x \longrightarrow 0$ then $t \longrightarrow 0$. We now have

$$\lim_{x \to 0} \frac{a^x - 1}{x} = \lim_{t \to 0} \frac{t}{\log_a (1 + t)} = \frac{1}{\log_a e} = \log a$$

and a special case is

$$\boxed{\lim_{x \to 0} \frac{e^x - 1}{x} = \log e = 1}$$

3.
$$\boxed{\lim_{x \to 0} \frac{(1 + x)^u - 1}{x} = u}$$

Proof: let $e^t = (1 + x)^u$; it follows $x = e^{\frac{t}{u}} - 1$ and $t = \log (1 + x)^u$ and we have:

$$\lim_{x \to 0} \frac{(1 + x)^u - 1}{x} = \lim_{t \to 0} \frac{e^t - 1}{e^{\frac{t}{u}} - 1} = \lim_{t \to 0} \frac{\frac{e^t - 1}{t}}{\frac{e^{\frac{t}{u}} - 1}{u \frac{t}{u}}} = \frac{1}{\frac{1}{u}} = u$$

4.
$$\boxed{\lim_{x \to 0} \frac{1 - \cos x}{x^2} = \frac{1}{2}}$$

Proof: $\lim_{x \to 0} \frac{1 - \cos x}{x^2} = \lim_{x \to 0} \frac{1 - \cos x}{x^2} \cdot \frac{1 + \cos x}{1 + \cos x} = \lim_{x \to 0} \frac{1 - (\cos x)^2}{x^2 (1 + \cos x)} =$

$= \lim_{x \to 0} \frac{(\sin x)^2}{x^2 (1 + \cos x)} \cdot \frac{1}{(1 + \cos x)} = 1 \cdot \frac{1}{2}$.

5.
$$\boxed{\lim_{x \to 0} \frac{\tan x}{x} = 1}$$

Proof: $\lim_{x \to 0} \frac{\tan x}{x} = \lim_{x \to 0} \frac{\sin x}{x} \frac{1}{\cos x} = 1 \cdot 1 = 1$

6.
$$\boxed{\lim_{x\to 0}\frac{\arctan x}{x}=1}$$

Proof: let $\arctan x = t$ then $x = \tan t$ with $t \longrightarrow 0$ when $x \longrightarrow 0$ hence:

$$\boxed{\lim_{x\to 0}\frac{\arctan x}{x}=\lim_{t\to 0}\frac{t}{\tan t}=1}$$

7.
$$\boxed{\lim_{x\to 0}\frac{\arcsin x}{x}=1}$$

proof: let $\arcsin x = t$ then $x = \sin t$ with $t \longrightarrow 0$ when $x \longrightarrow 0$ hence:

$$\boxed{\lim_{x\to 0}\frac{\arcsin x}{x}=\lim_{t\to 0}\frac{t}{\sin t}=1}$$

9.4 Asymptotic relations

Definition 9.2 *Let $c \in \overline{\mathbb{R}} = \mathbb{R} \cup \{-\infty, +\infty\}$;*

$f(x)$ is said infinite when $x \longrightarrow c$ if:

$$\lim_{x\to c} f(x) = \infty$$

$f(x)$ is said infinitesimal when $x \longrightarrow c$ if:

$$\lim_{x\to c} f(x) = 0$$

Now we consider three helpfull relations

1. **"Asymptothic equivalence"** relation; we say that $f(x)$ and $g(x)$ are asymptothically equivalent when $x \longrightarrow c$ if:

$$\lim_{x\to c}\frac{f(x)}{g(x)}=1$$

and we write $f(x) \sim g(x)$ when $x \longrightarrow c$.

Exercise 9.2 *(a) Prove that $\sim$ is an equivalence relation.*

(b) Determine if the following implications are true:

when $x \longrightarrow c$

$$\begin{cases} f(x) \sim h(x) \\ g(x) \sim h(x) \end{cases} \Longrightarrow [(f(x) + g(x)) \sim h(x)]$$

$$\begin{cases} f(x) \sim h(x) \\ g(x) \sim h(x) \end{cases} \Longrightarrow [(f(x) \cdot g(x)) \sim h(x)]$$

2. **"Same (order of) magnitude"**: we say that $f(x)$ and $g(x)$ have the same magnitude when $x \longrightarrow c$ if:

$$\lim_{x \to c} \frac{f(x)}{g(x)} = k \neq 0, k \in \mathbb{R}$$

and we write $f(x) \asymp g(x)$ when $x \longrightarrow c$.

Exercise 9.3 *Is $\asymp$ an equivalence?*

Are the following implications true?

$$[f(x) \sim h(x)] \Longrightarrow [f(x) \asymp h(x)]$$

$$[f(x) \asymp h(x)] \Longrightarrow [f(x) \sim h(x)]$$

$$\begin{cases} f(x) \asymp h(x) \\ g(x) \asymp h(x) \end{cases} \Longrightarrow [(f(x) + g(x)) \asymp h(x)]$$

$$\begin{cases} f(x) \asymp h(x) \\ g(x) \asymp h(x) \end{cases} \Longrightarrow [(f(x) \cdot g(x)) \asymp h(x)]$$

Definition 9.3 *if f and g are infinite (infinitesimal) when $x \longrightarrow c$ and $f(x) \asymp g(x)$ when $x \longrightarrow c$ we say they are infinite (infinitesimal) with the same (order of) magnitude.*

3. "Small o of": we say that $f(x)$ is a small o of $g(x)$ when $x \longrightarrow c$ if:

$$\lim_{x \to c} \frac{f(x)}{g(x)} = 0$$

and we write $f(x) = o(g(x))$ when $x \longrightarrow c$. We can also say that f is an infinitesimal of g and that f is negligible with respect to g when $x \longrightarrow c$.

Moreover:

if f and g are infinite when $x \longrightarrow c$ and $f(x) = o(g(x))$ we say that f is an infinite of smaller order (of magnitude) when $x \longrightarrow c$ with respect to g (g is an infinite of greater order);

if f and g are infinitesimal when $x \longrightarrow c$ and $f(x) = o(g(x))$ we say that f is an infinitesimal of greater order (of magnitude) when $x \longrightarrow c$ with respect to g (g is an infinitesimal of smaller order);

Exercise 9.4 *Prove that o is a strict partial order.*

Let $f(x) = \log x$ and $g(x) = x$, determine two real numbers c_1 and c_2 such that: $f(x) = o(g(x))$ when $x \longrightarrow c_1$ and $g(x) = o(f(x))$ when $x \longrightarrow c_2$.

Is the following implication true?

$$\begin{cases} f(x) = o(h(x)) \\ g(x) = o(h(x)) \end{cases} \implies [(f(x) + g(x)) = o(h(x))]$$

9.4.1 Sorting of the infinites

Small o is a partial weak order; hence it is possible to "sort" the infinites when $x \longrightarrow +\infty$ as follows:

$$x^x$$
$$a^x \quad \text{when } a > 1$$
$$x^w \quad \text{with } w \in \mathbb{R} \text{ and } w > 1$$
$$x$$
$$x^t \quad \text{with } t \in \mathbb{R} \text{ and } 0 < t < 1$$
$$\log_a x$$

Let's put it in other words: when $x \longrightarrow +\infty$:

$$\log_a x = o(x^t) \quad t > 0$$
$$x^t = o(x^w) \quad t < w$$
$$x^w = o(a^x) \quad a > 1$$
$$a^x = o(x^x)$$

Moreover if $a < b$ then $a^x = o(b^x)$ while $log_a x \asymp log_b x$.

Remark 9.2 *When $x \longrightarrow +\infty$ it is very natural to recognize little o, I mean it very easy to realize that in this case $x = o(x^{100})$. When $x \longrightarrow 0$ things are quite different: in this case $x^{100} = o(x)$! In general when $x \longrightarrow 0$ if $t < w$ then $x^w = o(x^t)$. Use numerical examples if you are not convinced.*

9.5 Elimination (reduction) principle

We can state the elimination principle as follows:

in the sum we can neglect the small o.

This means that if $h(x) = o(f(x))$ when $x \longrightarrow c$ then:

$$\lim_{x \to c} (f(x) + h(x)) = \lim_{x \to c} f(x)$$

Some authors present the elimination principle as follows:

if $h(x) = o(f(x))$ and $k(x) = o(g(x))$ when $x \longrightarrow c$ then:

$$\lim_{x \to c} \frac{f(x) + h(x)}{g(x) + k(x)} = \lim_{x \to c} \frac{f(x)}{g(x)}$$

Exercise 9.5 *prove the reduction principle.*

9.6 How to solve missing cases?

When facing a missing case there is not an universal algorithm we have to apply. Indeterminate cases can be solved in several ways and some techniques that works in many cases may be useless in other ones.

In the following I present some hint (they are hint not rules!).

- Apply the reduction principle and the order of the infinite for:

$$+\infty - \infty$$

$$\frac{\infty}{\infty}$$

$$0 \cdot \infty$$

Example 9.3

$$\lim_{x \to +\infty} \frac{2^x + \log x - 700x^4}{3^x - \sin x + \arctan x} = \lim_{x \to +\infty} \frac{2^x}{3^x} = 0$$

$$\lim_{x \to -\infty} xe^x = -\infty \cdot 0$$

Applying power properties and using the substitution $t = -x$ we may write:

$$\lim_{x \to -\infty} xe^x = \lim_{x \to -\infty} \frac{x}{e^{-x}} = \lim_{t \to +\infty} \frac{-t}{e^t} = 0$$

$$\lim_{x \to 0^+} x \log x = 0 \cdot (-\infty)$$

Using the substitution $t = \log x$ we may write:

$$\lim_{x \to 0^+} x \log x = \lim_{t \to -\infty} e^t t = 0$$

- Apply reduction principle, significant limits and factorization for:

$$\frac{0}{0}$$

Example 9.4

$$\lim_{x \to 0} \frac{x^2 + x}{x^2 - x} = \frac{0}{0}$$

Applying factorization we can write:

$$\lim_{x \to 0} \frac{x^2 + x}{x^2 - x} = \lim_{x \to 0} \frac{x(x+1)}{x(x-1)} = -1$$

Applying the reduction principle we can write:

$$\lim_{x \to 0} \frac{x^2 + x}{x^2 - x} = \lim_{x \to 0} \frac{x}{-x} = -1$$

$$\lim_{x \to 0} \frac{x \sin x + x}{x^2} = \frac{0}{0}$$

Writing in a different way the fraction and using one of the fundamental limits we obtain the solution:

$$\lim_{x \to 0} \frac{x \sin x - x}{x^2} = \lim_{x \to 0} \left[\frac{x \sin x}{x^2} - \frac{x}{x^2}\right] = \lim_{x \to 0} \left[\frac{\sin x}{x} - \frac{1}{x}\right] = 1 - \infty = \infty$$

Pay attention to last example: the result is ∞ not $-\infty$. Indeed we have:

$$\lim_{x \to 0^+} \left[\frac{\sin x}{x} - \frac{1}{x}\right] = 1 - \frac{1}{0^+} = -\infty$$

$$\lim_{x \to 0^-} \left[\frac{\sin x}{x} - \frac{1}{x}\right] = 1 - \frac{1}{0^-} = +\infty$$

- Reduction principle and significant limits for:

$$\infty^0$$

$$0^0$$

Example 9.5

$$\lim_{x \to +\infty} x^{\frac{1}{x}} = (+\infty)^{\frac{1}{+\infty}} = \infty^0$$

Recalling the rules concerning logarithms and exponentials we can write:

$$\lim_{x \to +\infty} x^{\frac{1}{x}} = \lim_{x \to +\infty} e^{\log\left[x^{\frac{1}{x}}\right]} = \lim_{x \to +\infty} e^{\frac{\log x}{x}} = e^0 = 1$$

$$\lim_{x \to 0^+} x^x = 0^0$$

$$\lim_{x \to 0^+} x^x = \lim_{x \to 0^+} e^{\log x^x} = \lim_{x \to 0^+} e^{x \log x} = e^0 = 1$$

- Significant limits for:

$$1^{+\infty}$$

Example 9.6

$$\lim_{x \to +\infty} \left(1 + \frac{1}{x}\right)^{x^2} = 1^{+\infty}$$

$$\lim_{x \to +\infty} \left(1 + \frac{1}{x}\right)^{x^2} = \lim_{x \to +\infty} \left[\left(1 + \frac{1}{x}\right)^x\right]^x = e^{+\infty} = +\infty$$

$$\lim_{x \to +\infty} \left(1 + \frac{1}{x^2}\right)^x = 1^{+\infty}$$

$$\lim_{x \to +\infty} \left(1 + \frac{1}{x^2}\right)^x = \lim_{x \to +\infty} \left[\left(1 + \frac{1}{x^2}\right)^{x^2}\right]^{\frac{1}{x}} = e^0 = 1$$

Chapter 10

Continuous functions

10.1 Basic definitions

Definition 10.1 *$f : A \subset \mathbb{R} \longrightarrow \mathbb{R}$ is said continuous in c if:*

$$\forall N\left(f\left(c\right)\right) \exists J\left(c\right) : x \in J\left(c\right) \Longrightarrow f\left(x\right) \in N\left(f\left(c\right)\right)$$

This definition is different by the limit definition because:

1. we consider $N(f(c))$ and this implies that c belongs to the domain of f; in the limit definition the function might not exist in c;

2. now $x \in J(c)$ and we don't have $x \neq c$ because in the limit definition we are not interested in the point c, but we are interested in what happens in a neighborhood of the point; dealing with continuous functions it is necessary that the function exists and has a right value in the point.

An equivalent definition is the following:

Definition 10.2 $f : A \subset \mathbb{R} \longrightarrow \mathbb{R}$ *is said continuous in c if:*

1. f *exists in c;*

2. $\lim\limits_{x \to c} f(x)$ *exists and is finite;*

3. $\lim\limits_{x \to c} f(x) = f(c)$.

The second condition above can be written as follows:

- $\lim\limits_{x \to c^+} f(x) = l_1$ (finite);

- $\lim\limits_{x \to c^-} f(x) = l_2$ (finite);

- $l_1 = l_2$.

We can use a third equivalent definition:

Definition 10.3 $f : A \subset \mathbb{R} \longrightarrow \mathbb{R}$ *is said continuous in c if:*

$$\lim\limits_{x \to c} f(x) = f(c)$$

Definition 10.4 $f : A \subset \mathbb{R} \longrightarrow \mathbb{R}$ *is said right continuous in c if:*

$$\lim\limits_{x \to c^+} f(x) = f(c)$$

$f : A \subset \mathbb{R} \longrightarrow \mathbb{R}$ *is said left continuous in c if:*

$$\lim\limits_{x \to c^-} f(x) = f(c)$$

Example 10.1 $f(x) = \sqrt{x}$ *is right continuous in 0 since*

$$\lim\limits_{x \to 0^+} \sqrt{x} = 0 = \sqrt{0}$$

Definition 10.5 *f is continuous in $]a;b[$ if f is continuous in every point of $]a;b[$;*

f is continuous in $[a;b]$ if f is continuous in every point of $]a;b[$, is right continuous in a and left continuous in b.

Proposition 10.1 *elementary functions are continuous in their domain. (the proof is not presented).*

Definition 10.6 *Given two functions f and g a linear combination of the two functions is:*

$$af(x) + bg(x) \qquad \text{where } a; b \in \mathbb{R}$$

Proposition 10.2 *sum, product, composition and linear combination of continuous functions are continuous functions in their domain. The proof is not presented.*

10.2 Discontinuities

Limit points of the domain of a function where the function is not continuous are called discontinuities; I use the following classification.

1. Jump discontinuity: c is a jump discontinuity of f if:

$$\lim_{x \to c^-} f(x) = l_1 \qquad \text{(finite)}$$
$$\lim_{x \to c^+} f(x) = l_2 \qquad \text{(finite)}$$
$$l_1 \neq l_2$$

$|l_1 - l_2|$ is called jump of the function.

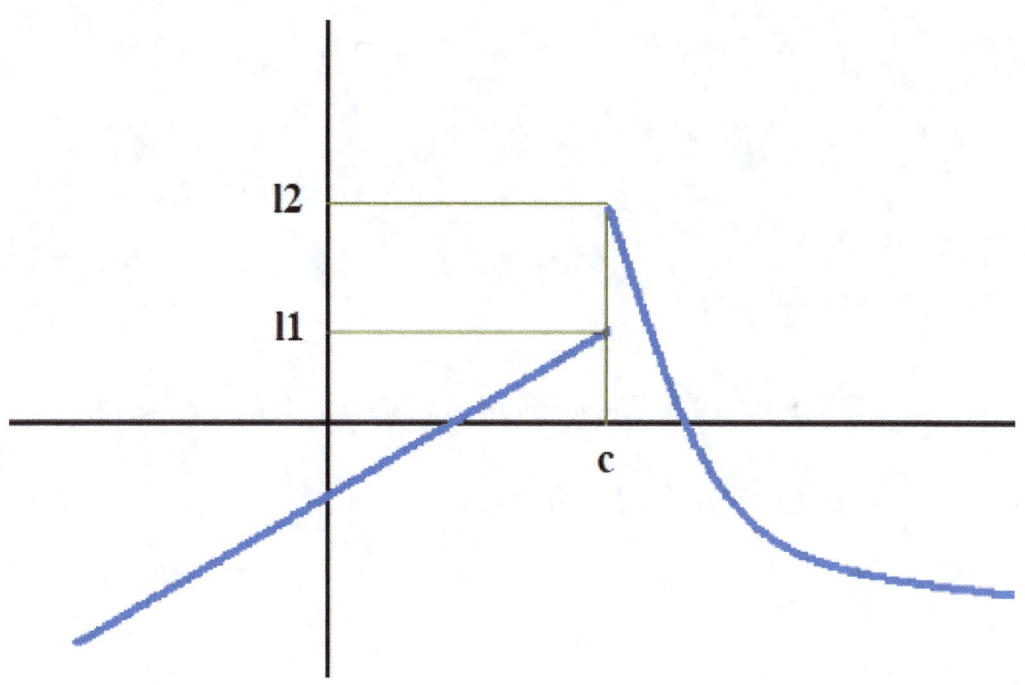

Example 10.2 *the function*

$$\text{sign}(x) = \begin{cases} +1 & \text{if } x > 0 \\ -1 & \text{if } x < 0 \end{cases}$$

is not continuous in 0 because it is not defined when $x = 0$;

and we can see that:

$$\lim_{x \to 0^-} f(x) = -1$$
$$\lim_{x \to 0^+} f(x) = +1$$

and the jump is 2.

Example 10.3 *The graph of the function $f(x) = \arctan \frac{1}{x}$ is*

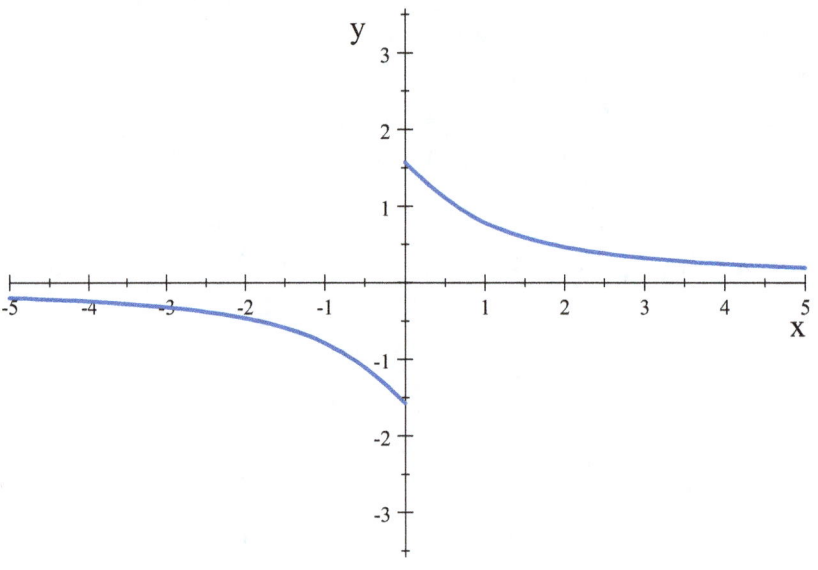

and

$$\lim_{x \to 0^-} \arctan \frac{1}{x} = \arctan(-\infty) = -\frac{\pi}{2}$$
$$\lim_{x \to 0^+} \arctan \frac{1}{x} = \arctan(+\infty) = +\frac{\pi}{2}$$

The jump is π.

Example 10.4 *I remind you that $floor(x) = \lfloor x \rfloor \stackrel{def}{=}$ the biggest integer number smaller or equal to x.*

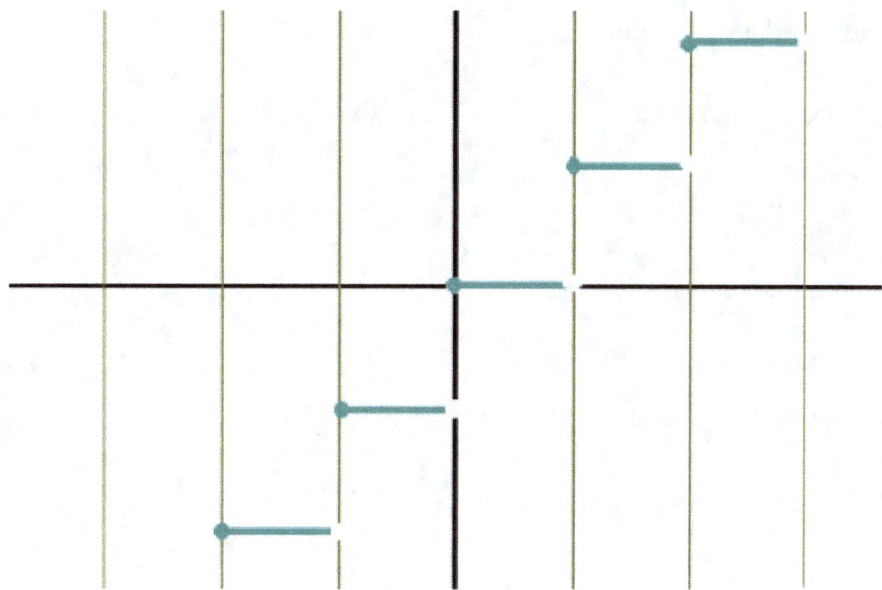

It has a jump whenever c is any integer number and the jump is 1; in these points the function is right continuous but it is not left continuous.

2. Infinite discontinuity:

   c is an infinite discontinuity if at least one between:

   $$\lim_{x \to c^-} f(x)$$

   $$\lim_{x \to c^+} f(x)$$

   is ∞.

   Some examples are $f(x) = \log_a x$ when $c = 0$ and $f(x) = \tan x$ when $c = \frac{\pi}{2}$.

   If c is an infinite discontinuity of f then the line $x = c$ is called vertical asymptote of the function; so $x = 0$ is vertical asymptote for logarithymic functions and $y = \frac{\pi}{2}$ is vertical asymptote for $\tan x$.

3. Essential discontinuity: c is an essential discontinuity if it is not an infinite one

and at least one between:

$$\lim_{x \to c^-} f(x)$$

$$\lim_{x \to c^+} f(x)$$

does not exists.

As an example let's think to the function $f(x) = sin\frac{1}{x}$ that we've already analyzed; it has an essential discontinuity in 0.

4. Removable discontinuity: c is removable discontinuity if:

$$\lim_{x \to c^-} f(x) = \lim_{x \to c^+} f(x) = l \text{ (finite)}$$

but $f(c)$ does not exist or $f(c) \neq l$. We can remove the discontinuity defining the new function (the e in the subscript stands for extended):

$$f_e(x) = \begin{cases} f(x) & \text{if } x \neq 0 \\ l & \text{if } x = 0 \end{cases}$$

Example 10.5

$$f(x) = \frac{x^2 - 3x + 2}{x - 1}$$

f is not continuous in 1 (the denominator cannot be equal to 0),

$$\lim_{x \to 1} \frac{x^2 - 3x + 2}{x - 1} = \lim_{x \to 1} \frac{(x-1)(x-2)}{x-1} = \lim_{x \to 1} (x - 2) = -1$$

we can write the function as follows:

$$f(x) = \begin{cases} x - 2 & \text{if } x \neq 1 \\ \text{does not exist} & \text{if } x = 1 \end{cases}$$

and a point is missing in its graph:

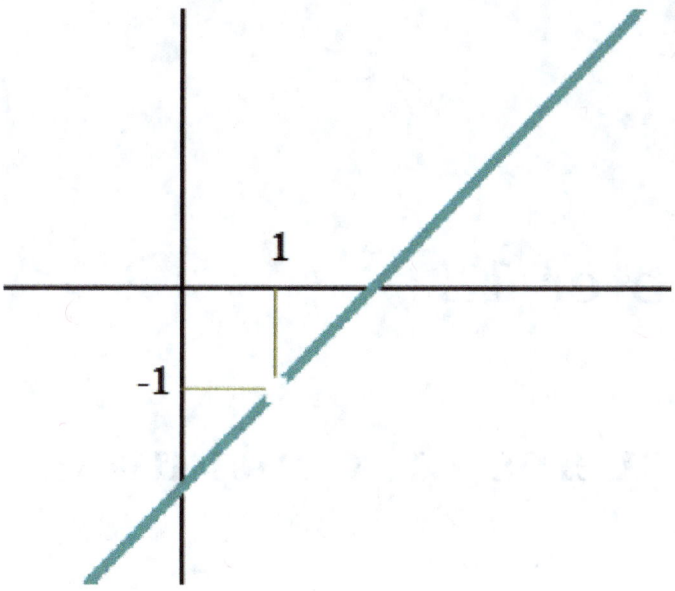

We can remove the discontinuity defining:

$$f_e(x) = \begin{cases} x - 2 & \text{if } x \neq 1 \\ -1 & \text{if } x = 1 \end{cases}$$

Example 10.6 *Let's come back to the function $f(x) = x \sin \frac{1}{x}$ (see page 119). $0 \notin D_f$ hence f is not continuous in $x = 0$ but we know that:*

$$\lim_{x \to 1} f(x) = 0$$

and we can remove the discontinuity defining the extended function:

$$f_e(x) = \begin{cases} x \sin \frac{1}{x} & \text{if } x \neq 0 \\ 0 & \text{if } x = 0 \end{cases}$$

The same happens for the function: $f(x) = x^2 \sin \frac{1}{x}$.

Exercise 10.1 *Determine and classify the discontinuities of the function:*

$f(a) :=$ *number of real solutions of the equation* $ax^2 - 3x + 2 = 0$

Chapter 11

Theorems on continuous functions.

11.1 Basic definitions:

$f(x)$ is strictly increasing in an ordered set A if $\forall a, b \in A$ we have

$$[a < b] \implies [f(a) < f(b)]$$

$f(x)$ is increasing in an ordered set A if $\forall a, b \in A$ we have

$$[a < b] \implies [f(a) \leq f(b)]$$

We have similar definitions for decreasing functions.

An (strictly) increasing or decreasing function is said (strictly) monotonic function.

The point c belonging to the domain of $f(x)$ is called global (absolute) maximum point if $f(c) \geq f(x) \forall x \in D_f$.

The point c belonging to the domain of $f(x)$ is called local (relative) maximum point if $\exists N(c) : f(c) \geq f(x) \forall x \in N(c)$.

We have similar definitions for minimum points.

11.2 Theorems

Proposition 11.1 *if f is continuous in $c \implies \exists J(c) : \forall x \in J(c) |f(x)| < +\infty$. In other word if f is continuous in c then it is bounded in a neighborhood of c.*

Proof. if f is continuous in c then

$$\forall N(f(c)) \exists J(c) : \forall x \in J(c), f(x) \in N(f(c)) \iff [(f(c) - \epsilon < f(x) < f(c) + \epsilon]$$

this means that the function is bounded above and below. ∎

Theorem 11.1 *(Sign theorem)*

$$\left. \begin{array}{c} f \text{ is continuous in } c \\ f(c) \neq 0 \end{array} \right\} \implies \exists J(c) : \forall x \in J(c), f(x) \cdot f(c) > 0$$

Proof. (we consider only the case $f(c) > 0$) if $f(x)$ is continuous in c then

$$\lim_{x \to c} f(x) = f(c) > 0$$

$$\Downarrow$$

$$\forall N(f(c)) \exists J(c) : x \in J(c)$$

$$\Downarrow$$

$$f(x) \in N(f(c))$$

The proof follows choosing $N(f(c))$ fully contained in in the set of the real positive numbers. ∎

Theorem 11.2 *(existence of the null values)*

$$\left. \begin{array}{c} f \text{ is continuous in } [a,b] \\ f(a) \cdot f(b) < 0 \end{array} \right\} \implies \exists c \in]a,b[: f(c) = 0$$

Proof. Let's consider $f(a) < 0$ (when $f(a) > 0$ almost everything is the same). Let $E := \{x : f(x) < 0\}$. We proove by contradiction that $f(c) = 0$ where $c = sup(E))$.

Let's suppose that $f(c) \neq 0$; this implies that either $f(c) > 0$ or $f(c) < 0$.

Let's assume $(f(c) < 0)$ then, by the sign theorem, $\exists J(c) : \forall x \in J(c), f(x) < 0$ and this implies that c cannot be the supremum of E since in the right of c there are points where f is negative;

let's now assume $(f(c) > 0)$ then, by the sign theorem, $\exists J(c) : \forall x \in J(c), f(x) > 0$ and this implies that c cannot be the supremum of E since in the left of c there are points where f is negative.

So $f(c)$ cannot be neither positive nor negative; conclusion: if $c = sup(E)$ then $f(c) = 0$. ∎

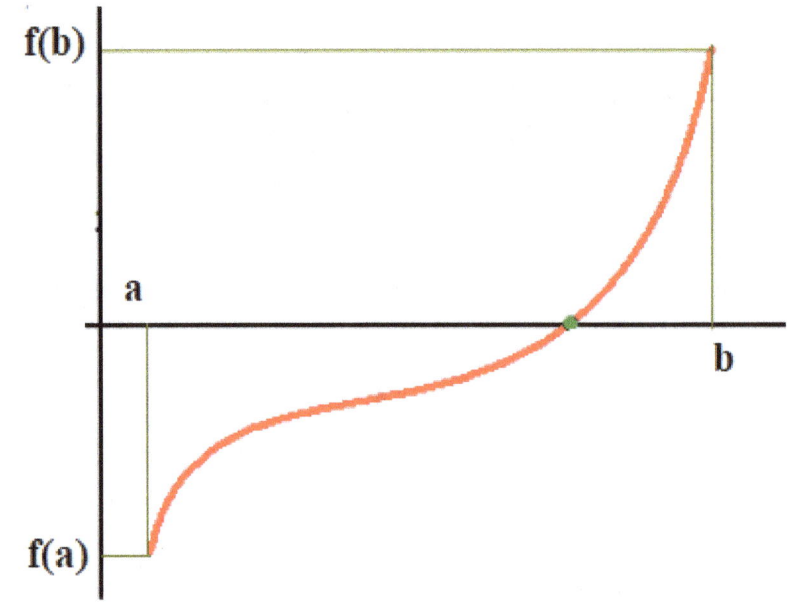

Theorem 11.3 *(Generalization of the theorem of existence of the null values)*

$$\left.\begin{array}{c} f \text{ is continuous in } \mathbb{R} \\ \left[\lim_{x \to -\infty} f(x)\right] \cdot \left[\lim_{x \to +\infty} f(x)\right] < 0 \end{array}\right\} \Longrightarrow \exists c \in \mathbb{R} : f(c) = 0$$

A further generalization of the theorem is the following:

Theorem 11.4 *(Intermediate values theorem or Darboux th.)*

$$\left.\begin{array}{c} f \text{ is continuous in } [a,b] \\ f(a) < f(b) \end{array}\right\} \Longrightarrow \forall \gamma \in]f(a), f(b)[\exists c \in]a,b[: f(c) = \gamma$$

that is $f(x)$ takes all the values between $f(a)$ and $f(b)$. The theorem works also in the case $f(a) > f(b)$.

Proof. Let us define the function

$$f_\gamma(x) = f(x) - \gamma$$

it is easy to realize that $f_\gamma(x)$ is continuous in $[a;b]$ since it is sum (linear combination) of continuous functions; moreover

$$\begin{cases} f_\gamma(a) = f(a) - \gamma < 0 \\ f_\gamma(b) = f(b) - \gamma > 0 \end{cases}$$

and by the theorem of null values $\exists c \in]a,b[: f_\gamma(c) = 0$ and this means that $f(c) = \gamma$. We can repeat this procedure for every γ and so the proof is complete. ∎

Theorem 11.5 *(Weiestrass Th.)*

if $f(x)$ is continuous in $[a;b]$ then it has global maximum and minimum (often in the boundary points!).

Theorem 11.6 *(Corollary of the Weierstrass Theorem)*

if $f(x)$ is continuous in $[a; b]$ then it takes all the values between minimum and maximum.

Proposition 11.2 *(existence of the inverse function).*

If f is continuous in $[a; b]$ and f^{-1} exists then f is strictly monotonic.

If a continuous function is not strictly monotonic it cannot be injective.

Exercise 11.1 Consider the equation: $e^x + x = 0$; has it solutions? Which theorem did you apply? Is the solution unique?

11.3 Transformations of continuous functions

Vertical translations

The graph of the function $f(x) = g(x) + k$ is the same of that of $g(x)$ but it is vertically moved (translated) (up if $k > 0$, down if $k < 0$).

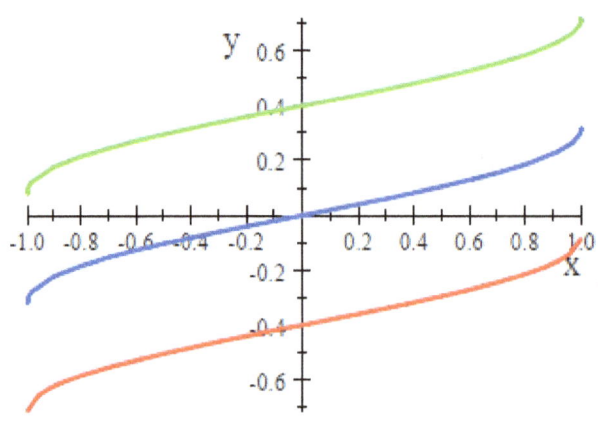

Horizontal translations

The graph of the function $f(x) = g(x - k)$ is the same of that of $g(x)$ but it is horizontally moved (translated) (to the left if $k > 0$, to the right if $k < 0$).

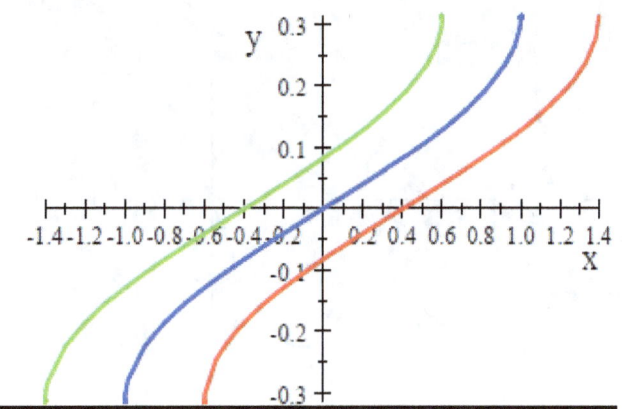

Vertical expansions and vertical contractions

Let $k > 0$ then the graph of the function $f(x) = kg(x)$ is the same of that of $g(x)$ but its is vertically expanded (when $k > 1$) or contracted (when $0 < k < 1$) (null values of $g(x)$ and $f(x)$ are the same).

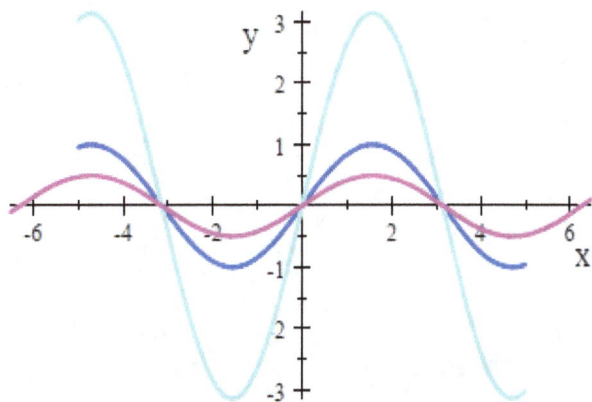

the functions presented are: $\sin x$, $\sin 2x$ and $\sin\left(\frac{1}{2}x\right)$.

Horizontal expansions and horizontal contractions

The graph of the function $f(x) = g(kx)$ is the same of that of $g(x)$ but it is horizontally expanded (when $0 < k < 1$) or contracted (when $k > 1$) (extrema of $g(x)$ and $f(x)$ are the same).

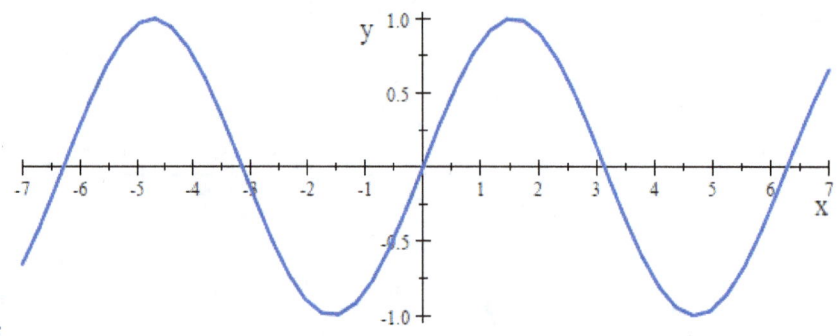

sin x

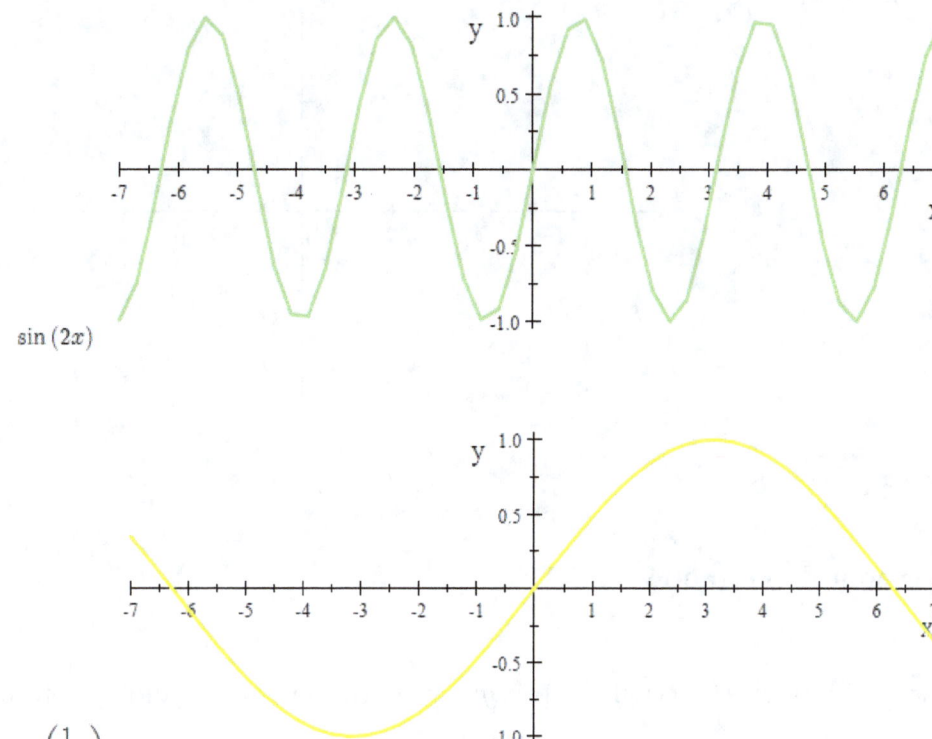

sin$(2x)$

sin$\left(\frac{1}{2}x\right)$

Vertical symmetry

The graph of the function $f = -g$ is the same of g and is obtained by symmetry with rispect to the horizontal axis (the null values of the two functions are the same, maxima becomes minima and viceversa).

In the graph we have: $g(x) = x^2 + 3x - 4$ (cyan) and $f(x) = -(x^2 + 3x - 4)$ (magenta).

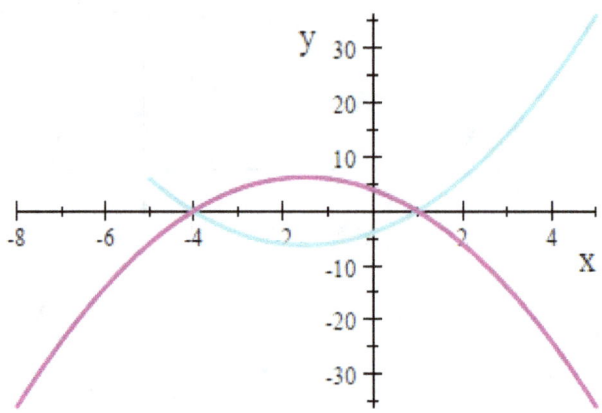

Horizontal symmetry

The graph of the function $f(x) = g(-x)$ is the same of g and is obtained by symmetry with rispect to the vertical axis. In the picture we have x^3 (cyan) and $(-x)^3$ (magenta)

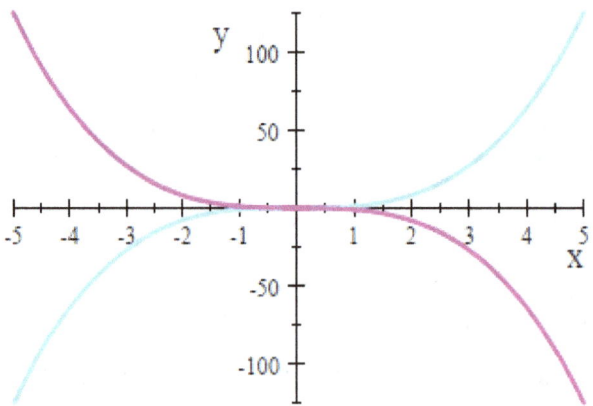

In next picture $\log x$ and $\log(-x)$ are represented.

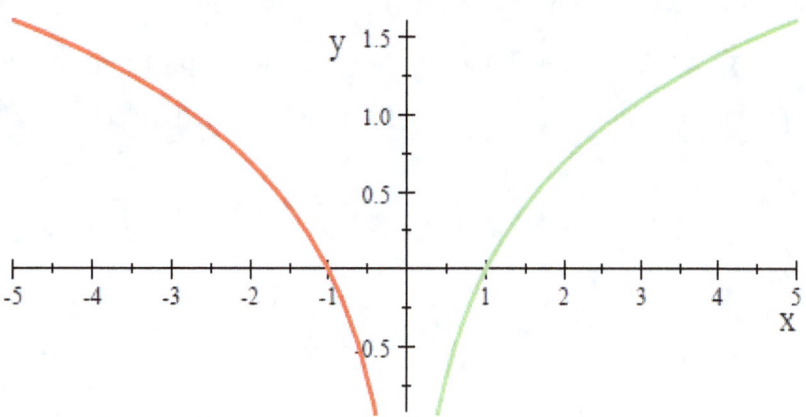

Effect of the absolute value

The graph of $|f(x)|$ can be obtained using the graph of $f(x)$: the graph remains the same when f is above the horizontal axis, become positive (by a vertical simmetry) when f is below the axis. In next picture we have $y = x^2 - 2x$ (dashed) and $y = |x^2 - 2x|$ (solid and thick).

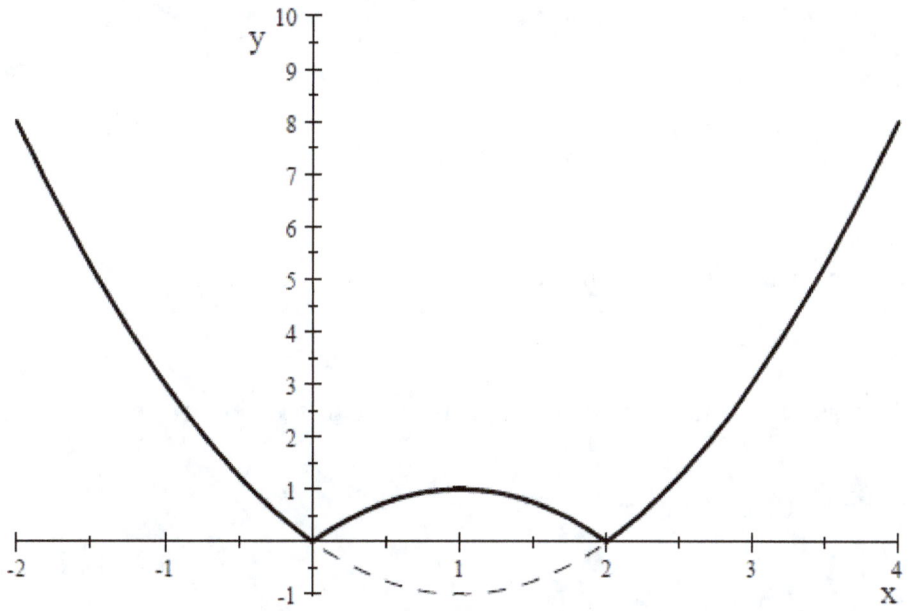

The graph of $f(|x|)$ can be obtained using the graph of $f(x)$: the graph remains the same when $x > 0$, and when $x < 0$ we get the graph by an horizontal simmetry. In next picture we have $f(x) = x^2 - 2x$ (dashed) and $f(|x|) = |x|^2 - 2|x|$ (solid and thick).

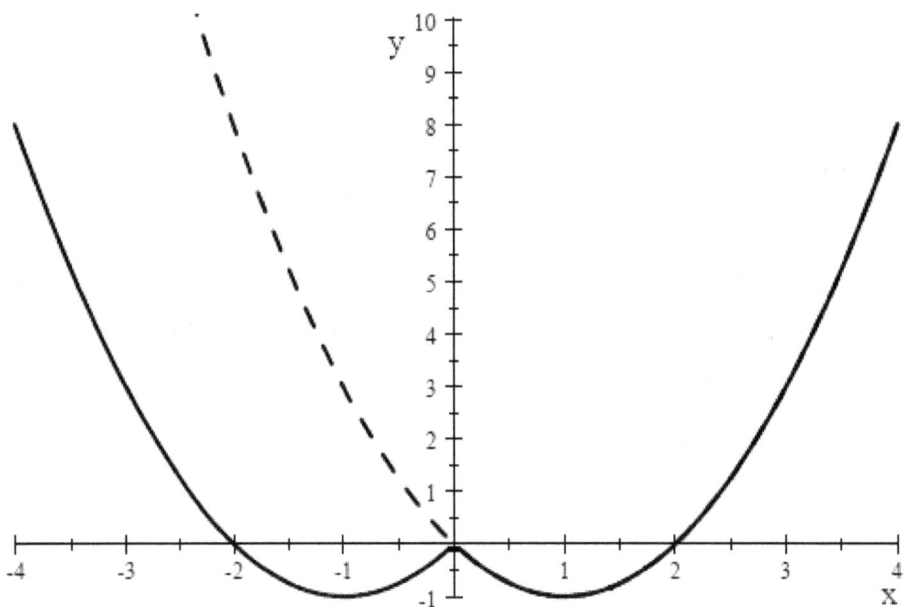

Compounding transformations

Just as an example in this paragraph we can see the graphs of the functions:

$\sin(\pi \cos(2x))$ (blue) and

$4\sin(cos(2x))$ (magenta).

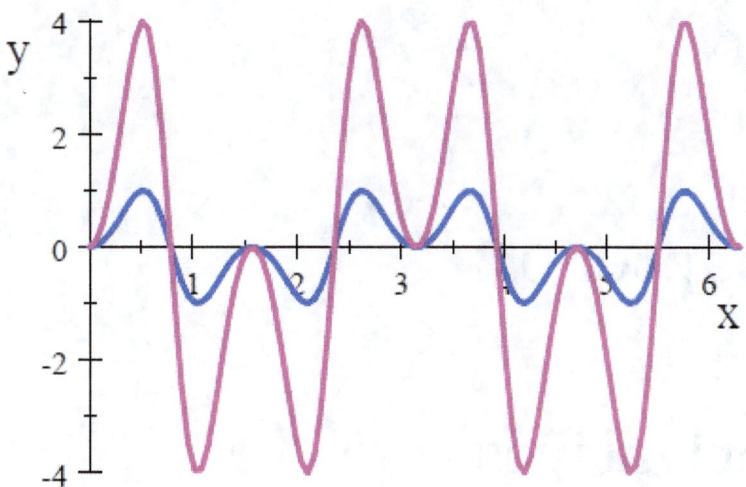

Chapter 12

Derivatives

12.1 Definition of derivative

Let $y = f(x)$ be a function $A \subseteq \mathbb{R} \longrightarrow \mathbb{R}$; the (first) derivative of f in c is the limit:

$$\lim_{\Delta x \to 0} \frac{\Delta y}{\Delta x} = \lim_{\Delta x \to 0} \frac{f(c + \Delta x) - f(c)}{\Delta x}$$

if this limit exists and is finite; Δx is the change of x.

If such a limit exists and is finite then the function is said smooth (or differentiable) in c; if $f(x)$ is smooth in all the points of a set A then $f(x)$ is smooth in A. Mathematicians denote the derivative in several ways; widely used notations are $f'(c)$, $Df|_c$, $\frac{\partial f}{\partial x}$, $\frac{df}{dx}$, $y'(c)$ and $\dot{y}(c)$.

The ratio $\dfrac{f(c + \Delta x) - f(c)}{\Delta x}$ is called difference ratio or difference quotient and is the slope of the secant line through the points $(c; f(c))$ and $(c + \Delta x; f(c + \Delta x))$:

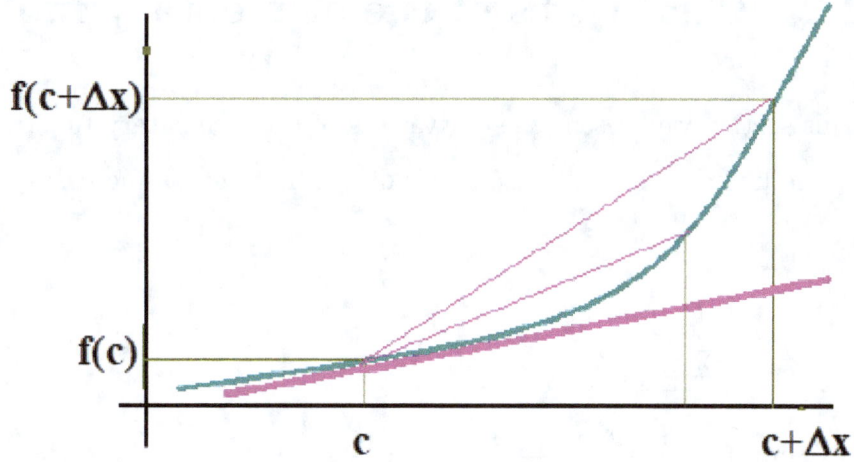

The limit in the definition can also be written:

$$\lim_{x \to c} \frac{f(x) - f(c)}{x - c}$$

where $(x - c) = x$.

Both way of writing are very important.

Geometric interpretation: the derivative in the point c is the slope of the tangent line to the graph of the function in the point $(c, f(c))$ (look at the picture above).

Definition 12.1 *The right derivative: is the limit:*

$$\lim_{\Delta x \to 0^+} \frac{f(c + \Delta x) - f(c)}{\Delta x}$$

If such a limit exists and is finite. The function is then right differentiable in c; we have a similar definition for the left derivative.

Up to now we have been considering the derivative in a fixed point c. If we consider the derivative in a variable point x then we are defining the derivative as a function and we write $f'(x)$ or $Df(x)$ or $(f(x))'$ or $\frac{\partial f(x)}{\partial x}$ and $\frac{df(x)}{dx}$.

12.2 Derivatives of the elementary functions

In this section we present the derivatives of the elementary functions; keep in mind that the limits we are going to solve always involve missing cases.

1.
$$\boxed{Dk = 0}$$
$$\lim_{\Delta x \to 0} \frac{f(x + \Delta x) - f(x)}{\Delta x} = \lim_{\Delta x \to 0} \frac{k - k}{\Delta x} = 0.$$

2.
$$\boxed{Dx = 1}$$
$$\lim_{\Delta x \to 0} \frac{f(x + \Delta x) - f(x)}{\Delta x} = \lim_{\Delta x \to 0} \frac{x + \Delta x - x}{\Delta x} = 1.$$

3.
$$\boxed{Dx^n = nx^{n-1}}$$

if $n \in \mathbb{N}$; $n > 1$ we can prove the above equality the proof using the binomial coefficients (not covered in this book)

$$\lim_{\Delta x \to 0} \frac{f(x + \Delta x) - f(x)}{\Delta x} = \lim_{\Delta x \to 0} \frac{(x + \Delta x)^n - x^n}{\Delta x} =$$
$$= \lim_{\Delta x \to 0} \frac{x^n + \binom{n}{1} x^{n-1} \Delta x + \binom{n}{2} x^{n-2} (\Delta x)^2 + \dots + (\Delta x)^n - x^n}{\Delta x}$$

Applying the elimination principle we can write:
$$\lim_{\Delta x \to 0} \frac{x^n + \binom{n}{1} x^{n-1} \Delta x + \binom{n}{2} x^{n-2} (\Delta x)^2 + \dots + (\Delta x)^n - x^n}{\Delta x} =$$
$$= \lim_{\Delta x \to 0} \frac{n x^{n-1} \Delta x}{\Delta x} = n x^{n-1}$$

In general if $n \in \mathbb{R}$ we have:
$$\lim_{\Delta x \to 0} \frac{(x + \Delta x)^n - x^n}{\Delta x} = \lim_{\Delta x \to 0} x^n \frac{\left(1 + \frac{\Delta x}{x}\right)^n - 1}{\Delta x} = \lim_{\Delta x \to 0} \frac{x^n}{x} \frac{\left(1 + \frac{\Delta x}{x}\right)^n - 1}{\frac{\Delta x}{x}}$$

and applying the right significant limit we have

$$\lim_{\Delta x \to 0} \frac{x^n}{x} \frac{\left(1 + \frac{\Delta x}{x}\right)^n - 1}{\frac{\Delta x}{x}} = nx^{n-1}$$

4.

$$\boxed{Da^x = a^x \log a}$$

$$\lim_{\Delta x \to 0} \frac{f(x + \Delta x) - f(x)}{\Delta x} = \lim_{\Delta x \to 0} \frac{a^{x+\Delta x} - a^x}{\Delta x} = \lim_{\Delta x \to 0} a^x \frac{a^{\Delta x} - 1}{\Delta x} = a^x \log a.$$

Since $\log e = 1$ we have

$$\boxed{De^x = e^x}$$

This is one of the nice properties of e we spoke about at page 132.

5.

$$\boxed{D \log_a x = \frac{1}{x} \log_a e}$$

$$\lim_{\Delta x \to 0} \frac{f(x + \Delta x) - f(x)}{\Delta x} = \lim_{\Delta x \to 0} \frac{\log_a(x + \Delta x) - \log_a x}{\Delta x} =$$

$$= \lim_{\Delta x \to 0} \frac{\log_a \frac{x + \Delta x}{x}}{\Delta x} = \lim_{\Delta x \to 0} \frac{1}{\Delta x} \log_a \left(1 + \frac{\Delta x}{x}\right) =$$

$$= \lim_{\Delta x \to 0} \frac{1}{x} \log_a \left(1 + \frac{\Delta x}{x}\right)^{\frac{x}{\Delta x}} = \frac{1}{x} \log_a e.$$

And we have

$$\boxed{D \log x = \frac{1}{x}}$$

This is another one of the nice properties of e we spoke about.

6.

$$\boxed{D \sin x = \cos x}$$

First of all let me remind you the trigonometric identity

$$\sin(\alpha + \beta) = \sin\alpha \cos\beta + \sin\beta \cos\alpha$$

$$\lim_{\Delta x \to 0} \frac{f(x+\Delta x) - f(x)}{\Delta x} = \lim_{\Delta x \to 0} \frac{\sin(x+\Delta x) - \sin x}{\Delta x} =$$

$$\lim_{\Delta x \to 0} \frac{\sin x \cdot \cos \Delta x + \sin \Delta x \cdot \cos x - \sin x}{\Delta x} =$$

$$= \lim_{\Delta x \to 0} \left[\frac{\sin x \cdot \cos \Delta x - \sin x}{\Delta x} + \frac{\sin \Delta x \cdot \cos x}{\Delta x} \right] =$$

$$= \lim_{\Delta x \to 0} \left[\sin x \frac{\cos \Delta x - 1}{(\Delta x)^2} \Delta x + \cos x \frac{\sin \Delta x}{\Delta x} \right] = \cos x.$$

7.

$$\boxed{D \cos x = -\sin x}$$

Remember that

$$\cos(\alpha + \beta) = \cos\alpha \cos\beta - \sin\alpha \cos\beta$$

$$\lim_{\Delta x \to 0} \frac{f(x+\Delta x) - f(x)}{\Delta x} = \lim_{\Delta x \to 0} \frac{\cos(x+\Delta x) - \cos x}{\Delta x} =$$

$$= \lim_{\Delta x \to 0} \frac{\cos x \cdot \cos \Delta x - \sin \Delta x \cdot \sin x - \cos x}{\Delta x} =$$

$$= \lim_{\Delta x \to 0} \left[\frac{\cos x \cdot \cos \Delta x - \cos x}{\Delta x} - \frac{\sin \Delta x \cdot \sin x}{\Delta x} \right] =$$

$$= \lim_{\Delta x \to 0} \left[\cos x \frac{\cos \Delta x - 1}{(\Delta x)^2} \Delta x - \sin x \frac{\sin \Delta x}{\Delta x} \right] = -\sin x.$$

12.3 Computation Rules

Derivative of a product

$$\boxed{D[f(x) \cdot g(x)] = f'(x) \cdot g(x) + f(x) \cdot g'(x)}$$

$$D[f(x)\cdot g(x)] = \lim_{\Delta x \to 0} \frac{f(x+\Delta x)\cdot g(x+\Delta x) - f(x)\cdot g(x)}{\Delta x} =$$

$$= \lim_{\Delta x \to 0} \frac{f(x+\Delta x)g(x+\Delta x) - f(x)g(x+\Delta x) + f(x)g(x+\Delta x) - f(x)g(x)}{\Delta x} =$$

$$= \lim_{\Delta x \to 0} \left[g(x+\Delta x)\frac{f(x+\Delta x) - f(x)}{\Delta x} + f(x)\frac{g(x+\Delta x) - g(x)}{\Delta x} \right] =$$

$$= f'(x)\cdot g(x) + f(x)\cdot g'(x)$$

Multiplying constant (remains unchanged)

$$\boxed{D[k\cdot f(x)] = kf'(x)}$$

$$D[k\cdot f(x)] = 0\cdot f(x) + kf'(x)$$

Ratio

$$\boxed{D[f(x)\cdot g(x)] = \frac{f'(x)\cdot g(x) - f(x)\cdot g'(x)}{[gx]^2}}$$

The proof is not shown.

Example 12.1

$$D\tan x = \frac{1}{\cos^2 x} = 1 + \tan^2 x$$

$$D\frac{\sin x}{\cos x} = \frac{\cos x \cdot \cos x - \sin x \cdot (-\sin x)}{(\cos x)^2} = \frac{1}{\cos^2 x} = 1 + \tan^2 x$$

Exercise 12.1 *Compute the derivative of* $\cot x = \frac{\cos x}{\sin x}$

Chain rule

$$\boxed{D[f(g(x))] = f'(g(x))\cdot g'(x)}$$

$$D[f(g(x))] = \lim_{\Delta x \to 0} \frac{f(g(x+\Delta x)) - f(g(x))}{\Delta x} =$$

$$= \lim_{\Delta x \to 0} \frac{f(g(x+\Delta x)) - f(g(x))}{g(x+\Delta x) - g(x)} \cdot \frac{g(x+\Delta x) - g(x)}{\Delta x} = f'(g(x)) \cdot g'(x)$$

Inverse function (1)

If the tangent line to the graph of $f(x)$ in the point $(c; f(c))$ has an angle α with the horizontal axis; the tangent line to the graph of the inverse function in the point $(f(c); c)$ has an angle $(90° - \alpha)$ degrees. Since:

$$\tan \alpha = \frac{1}{\tan(90° - \alpha)}$$

it follows:

$$Df^{-1}(f(c)) = \frac{1}{f'(c)}$$

$$f(x) = x^3 \qquad f'(x) = 3x^2 \qquad c = 2 \qquad f'(c) = 12$$
$$f^{-1}(x) = \sqrt[3]{x} \qquad Df^{-1}(x) = \frac{1}{3\sqrt[3]{x^2}} \qquad f(c) = 8 \qquad Df^{-1}(f(c)) = \frac{1}{12}$$

Inverse function (2)

We can compute the derivative of the inverse function also by the difference ratio:

let $y = f(x)$ and $y_c = f(c)$;

$$\lim_{y \to y_c} \frac{f^{-1}(y) - f^{-1}(y_c)}{y - y_c} = \lim_{y \to y_c} \frac{x - c}{f(x) - f(c)} = \frac{1}{f'(c)}$$

Derivatives of the inverse ot the trigonometric functions

$$\boxed{D \arctan x = \frac{1}{1+x^2}}$$

Let $y = \arctan x$, hence $x = \tan y$;

$$D[\arctan(x)] = \frac{1}{D\tan y} = \frac{1}{1+\tan^2 y} = \frac{1}{1+[\tan(\arctan x)]^2} = \frac{1}{1+x^2}$$

$$\boxed{D\arcsin x = \frac{1}{\sqrt{1-x^2}}}$$

Let $y = \arcsin x$, hence $x = \sin y$;

$$D[\arctan(x)] = \frac{1}{D\sin y} = \frac{1}{\cos y} = \frac{1}{\sqrt{1-\sin^2 y}} = \frac{1}{\sqrt{1-\sin^2(\arcsin x)}} = \frac{1}{\sqrt{1-x^2}}$$

In the same way we can proove that:

$$\boxed{D\operatorname{arccot} x = -\frac{1}{1+x^2}}$$

$$\boxed{D\arccos x = -\frac{1}{\sqrt{1-x^2}}}$$

12.4 Points where a continuous function is not smooth

Definition 12.2 *the point c is a kink if:*

1. $f(x)$ *is continuous in c;*

2. $\lim_{x \to c^+} f'(x) = l_1$ *(finite);*

3. $\lim_{x \to c^-} f'(x) = l_2$ *(finite);*

4. $l_1 \neq l_2$.

Example 12.2 *The function $y = |x|$ has a kink in 0; in fact*

$$f(x) = \begin{cases} +x & \text{if } x \geq 0 \\ -x & \text{if } x \leq 0 \end{cases}$$

$$f'(x) = \begin{cases} +1 & \text{if } x > 0 \\ -1 & \text{if } x < 0 \end{cases}$$

$y = |\sin x|$ *(in the set $[0; 2\pi]$) has a kink in π:*

$$f(x) = \begin{cases} +\sin x & \text{if } x > \pi \\ -\sin x & \text{if } x < \pi \end{cases}$$

$$f'(x) = \begin{cases} +\cos x & \text{if } x > \pi \\ -\cos x & \text{if } x < \pi \end{cases}$$

$$\lim_{x \to \pi^+} f'(x) = 1$$

$$\lim_{x \to \pi^-} f'(x) = -1$$

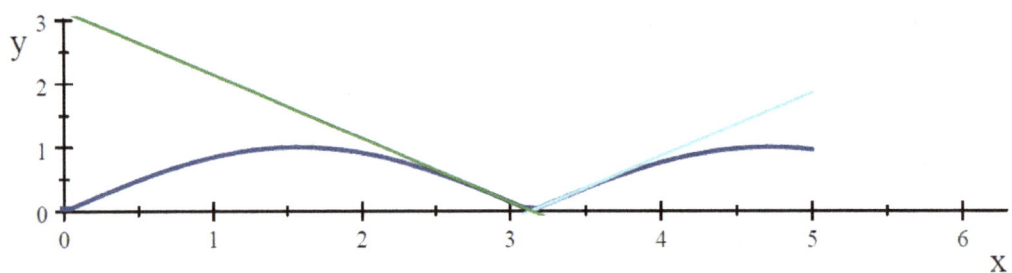

Example 12.3 *The function $y = |\log x|$ has a kink point in $x = 1$:*

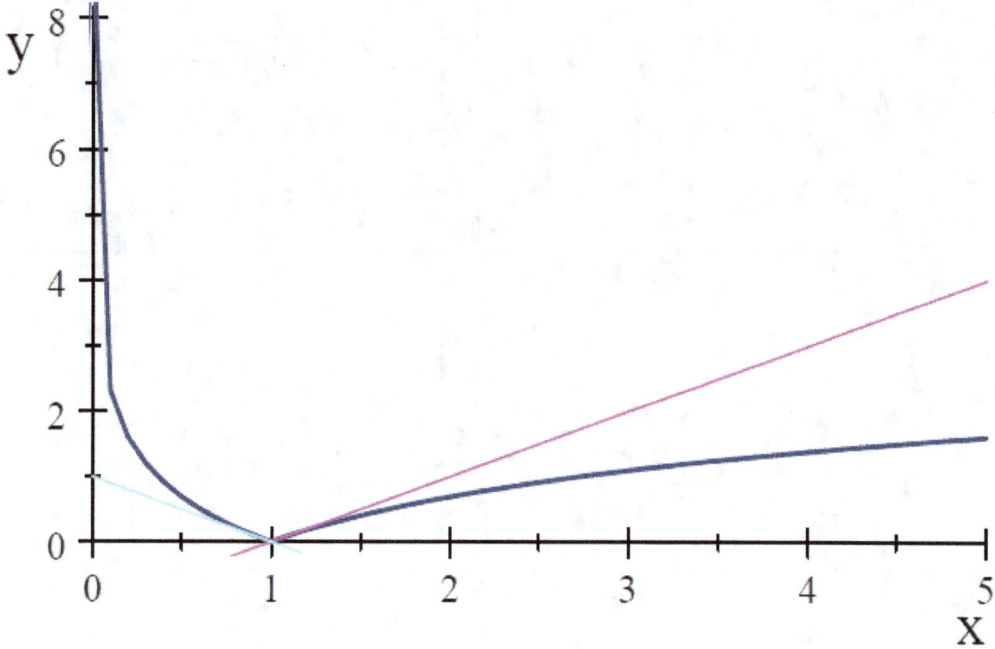

Definition 12.3 *The point c is a turning (inflection) point with vertical tangent line if:*

1. $f(x)$ is continuous in c;

2. $\lim\limits_{x \to c^+} f'(x) = +\infty \ (-\infty)$;

3. $\lim\limits_{x \to c^-} f'(x) = l_2 +\infty \ (-\infty)$;

Example 12.4 *Functions with vertical tangent are $y = \sqrt[3]{x}$ when $x = 0$ and $y = \sqrt[3]{x-1}$ when $x = 1$. The picture shows the graph of both functions (please notice the horizontal traslation):*

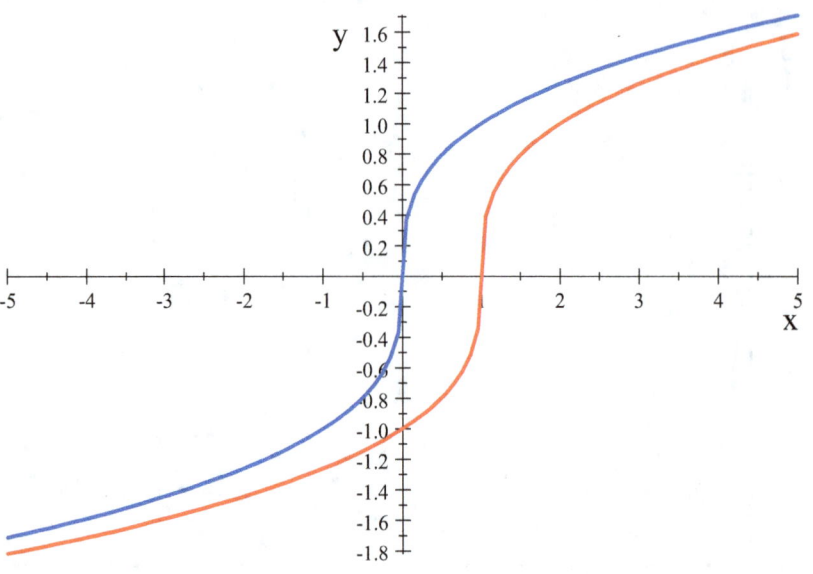

Definition 12.4 *the point c is a cusp if:*

1. $f(x)$ is continuous in c;

2. $\lim\limits_{x \to c^+} f'(x) = +\infty \ (-\infty)$;

3. $\lim\limits_{x \to c^-} f'(x) = -\infty \ (+\infty)$;

Example 12.5 *A function with a cusp in 1 is $y = \sqrt[3]{(x-1)^2}$; here is its graph:*

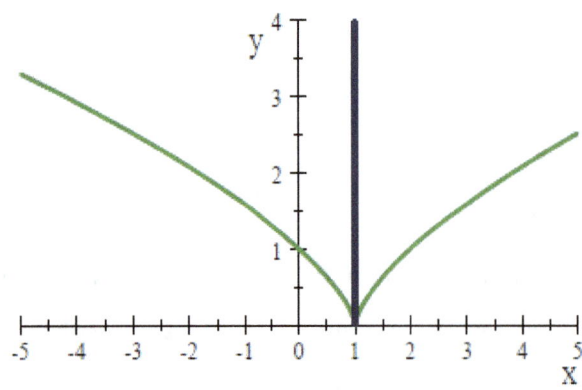

Definition 12.5 *c is a strange point if:*

1. *$f(x)$ is continuous in c;*

2. *At least one of the following limits does not exist:*

$$\lim_{x \to c^+} f'(x) \qquad \lim_{x \to c^-} f'(x)$$

Example 12.6 *the function*

$$y = \begin{cases} x \sin \frac{1}{x} & \text{if } x \neq 0 \\ 0 & \text{if } x = 0 \end{cases}$$

has a strange point in 0 (please check).

The function

$$y = \begin{cases} x^2 \sin \frac{1}{x} & \text{if } x \neq 0 \\ 0 & \text{if } x = 0 \end{cases}$$

has a strange point in 0 (please check).

The graphs of these functions are at pages 2.

12.5 Theorems on smooth functions

Proposition 12.1 *f is smooth in $c \implies f$ is continuous in c.*

Proof. f is smooth in $c \implies \lim_{\Delta x \to 0} \dfrac{f(c + \Delta x) - f(c)}{\Delta x} = l \in \mathbb{R} \implies$

$$\implies \begin{cases} [f(c + \Delta x) - f(c)] \asymp \Delta x \\ \qquad \text{or} \\ [f(c + \Delta x) - f(c)] \sim \Delta x \end{cases} \implies$$

$$\implies \lim_{\Delta x \to 0} f(c + \Delta x) - f(c) = 0 \implies \lim_{x \to c} f(x) = f(c). \blacksquare$$

Hence to be smooth is a sufficient condition to be continuous; moreover since the proposition is equivalent (via contradiction principle) to

$$\text{not continuous} \implies \text{not smooth}$$

to be continuous is a necessary condition to be smooth.

Theorem 12.1 *(Rolle th.)*

$$\begin{cases} f \text{ continuous in } [a,b] \\ f \text{ smooth in }]a,b[\\ f(a) = f(b) \end{cases} \implies \exists c \in]a,b[: f'(c) = 0$$

Proof. f is continuous in $[a,b] \implies$ (Weierstrass theorem) f has minimum and maximum in $[a,b]$; we have two possible cases:

greatest and smallest values are equal; hence the function is constant in $[a;b] \implies f'(c) = 0 \forall c \in]a;b[$;

the greatest value is greater than the smallest one $\implies$ at least one is in the interior of the interval $[a;b]$ (see picture).

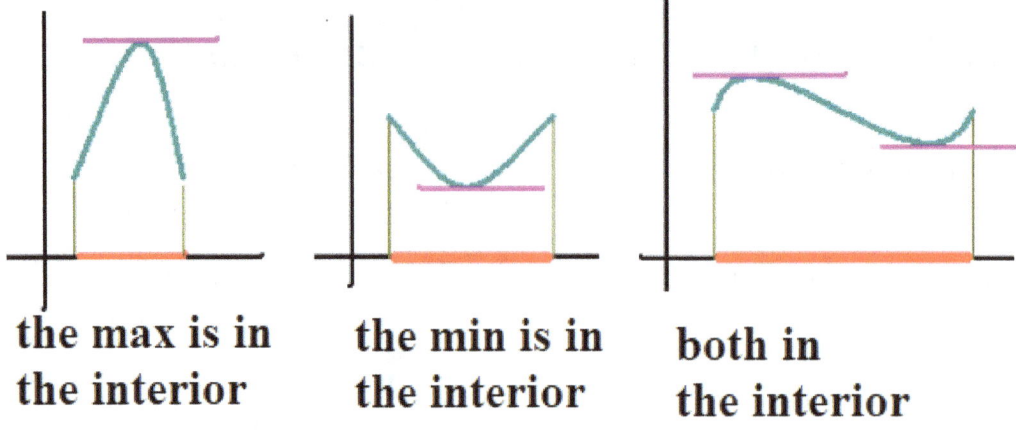

the max is in the interior **the min is in the interior** **both in the interior**

Let suppose the greatest is in the interior; let c be the maximum point:

$$c \in]a;b[\Longrightarrow \exists J(c) : f(c+\Delta x) \leq f(c)) \Longrightarrow$$

$$\Longrightarrow \begin{cases} \dfrac{f(c+\Delta x) - f(c)}{\Delta x} \leq 0 & \text{if} \quad \Delta x > 0 \\ \dfrac{f(c+\Delta x) - f(c)}{\Delta x} \geq 0 & \text{if} \quad \Delta x < 0 \end{cases} \Longrightarrow$$

$$\Longrightarrow \begin{cases} \lim\limits_{\Delta x \to 0^+} \dfrac{f(c+\Delta x) - f(c)}{\Delta x} \leq 0 \\ \lim\limits_{\Delta x \to 0^-} \dfrac{f(c+\Delta x) - f(c)}{\Delta x} \geq 0 \end{cases}$$

Since the function is smooth the two limits must be equal $\Longrightarrow f'(c) = 0$. ∎

Geometric interpretation

There exists at least a point where the tangent line to the graph of the function is horizontal.

Theorem 12.2 *(Lagrange[1] Th.)*

$$\begin{cases} f \text{ continuous in } [a,b] \\ f \text{ smooth in }]a,b[\end{cases} \Longrightarrow \exists c \in]a,b[: f'(c) = \frac{f(b) - f(a)}{b - a}$$

To prove the theorem we consider the function obtained by the difference between $f(x)$ and the line $r(x)$ through the points $(a; f(a))$ and $(b; f(b))$; this function satisfies the assumptions of Rolle theorem and this implies the consequence.

Proof. Let

$$r(x) = f(a) + \frac{f(b) - f(a)}{b - a}(x - a)$$

the points $(a; f(a))$ and $(b; f(b))$ belongs to this line then the function:

$$g(x) = f(x) - r(x)$$

satisfies the assumptions of the Rolle theorem since:

---

[1] Lagrange was an Italian mathematician born in Torino.

1. it is continuous in $[a;b]$ because it is sum of continuous functions;

2. it is smooth in $]a;b[$ because it is sum of smooth functions;

3. $g(a) = g(b)$ since:
$$g(a) = f(a) - \left[f(a) + \frac{f(b)-f(a)}{b-a}(a-a)\right] = 0$$
$$g(b) = f(b) - \left[f(a) + \frac{f(b)-f(a)}{b-a}(b-a)\right] = 0$$

hence $\exists c \in]a;b[: g'(c) = 0$ that is:
$$f'(c) - \frac{f(b)-f(a)}{b-a}(1-0) = 0$$
$$\Updownarrow$$
$$f'(c) = \frac{f(b)-f(a)}{b-a}$$

■

Geometric interpretation

There exists at least a point where the tangent line to the graph of the function is parallel to the secant line to the graph of the function; in fact the slope of the secant line is $\frac{f(b)-f(a)}{b-a}$ while the slope of the tangent line is $f'(c)$ and if the slope are the same the lines are parallel.

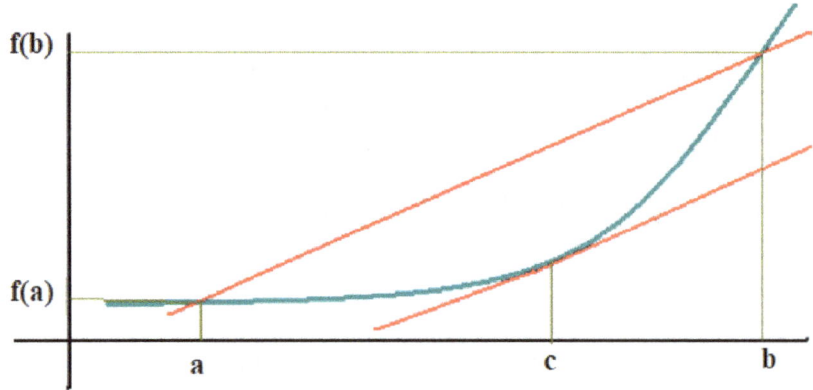

Remark 12.1 *the Rolle theorem is a special case of the Lagrange theorem.*

Corollary 12.1 *(first consequence of the Lagrange th):*

if the derivative of a function is always equal to zero then the function is constant, that is

$$f'(c) = 0 \forall c \in \,]a;b[\implies f(x) = k$$

Proof. $f'(c) = 0 \forall c \in \,]a;b[\implies f(x_1) = f(x_2) \forall \,]x_1;x_2[\subset \,]a;b[.$ ∎

Corollary 12.2 *(second consequence of the Lagrange th):*

if two functions have the same derivative everywhere then they differ for an additive constant; in other words this means that if $f(x)$ is obtained by a vertical translation of $g(x)$, then they have always the same derivative, that is:

$$[f'(c) = g'(c) \forall c] \implies [f(x) = g(x) + k]$$

Keep in mind that this corollary will be used while solving integrals.

Example 12.7 *Let's consider functions the $f(x) = \sin^2 x$ and $g(x) = -\cos^2 x$; we have*

$$D(\sin x)^2 = 2 \sin x \cos x$$
$$D(-\cos x)^2 = -2 \cos x (-\sin x) = 2 \sin x \cos x$$

The two functions have the same derivative since (because of the fundamental identity of the trigonometry)

$$\sin^2 x = -\cos^2 x + 1$$

Corollary 12.3 *(third consequence)*

$$f'(c) > 0 \forall c \in]a;b[$$
$$\Downarrow$$
$$\frac{f(b) - f(a)}{b - a} > 0$$
$$f(b) > f(a)$$

this means that if the derivative is positive (negative) in all the points of an interval then the function is increasing (decreasing) in that interval.

Example 12.8 *Let's consider the function $y = x^2$; its derivative, $y' = 2x$, is negative when $x < 0$ and positive when $x > 0$ and we can realize that the slope of the tangent line is negative when x is negative, positive when $x > 0$. When $x = 0$ the function has a minimum point and its derivative is null (see picture below).*

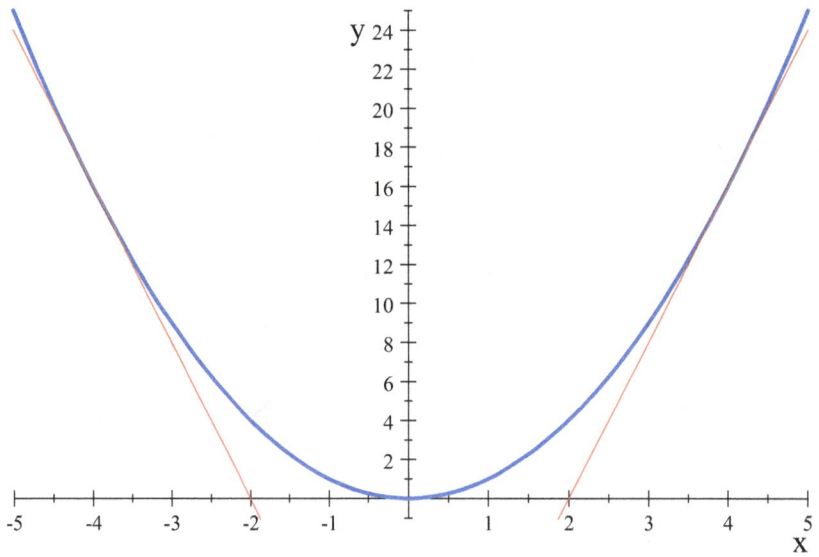

Exercise 12.2 *Identify a function and an interval $]a;b[$ satisfying the theorem in infinite many points belonging to $]a;b[$.*

Theorem 12.3 *De l'Hospital's rule*

$$\left.\begin{array}{l} f \text{ and } g \text{ are continuous in } N(c) - \{c\} \\ g'(c) \neq 0 \text{ in } N(c) - \{c\} \\ \lim_{x \to c} \dfrac{f(x)}{g(x)} = \dfrac{0}{0} \text{ (or } \dfrac{\infty}{\infty}) \\ \lim_{x \to c} \dfrac{f'(x)}{g'(x)} \text{ does exist} \end{array}\right\} \implies \lim_{x \to c} \dfrac{f(x)}{g(x)} = \lim_{x \to c} \dfrac{f'(x)}{g'(x)}$$

the proof is not presented.

12.5.1 Some notes on De L'Hopital rule

The rule is quite useful to solve some missing cases that we cannot solve using other technicques. The application of the rule is recommended in the following cases (the reader is asked to compute the limits):

$$\lim_{x \to 0} \frac{x - \sin x}{x^3} \qquad \lim_{x \to 0} \frac{\tan x - x}{x^3}$$
$$\lim_{x \to 0} \frac{x - \arcsin x}{x^3} \qquad \lim_{x \to 0} \frac{\arctan x - x}{x^3}$$

The above limits cannot be computed using significant limits or the elimintation principle and so applying De L'Hopital rule may be a winning strategy.

The limit

$$\lim_{x \to 0} \frac{\tan x - \sin x}{x^3}$$

can be solved using significant limits, using De L'Hopital rule or combining both methods (Please try).

1. The rule can be applied even if $c = \pm \infty$.

2. Never apply this rule when working with sequences (not covered in this book).

3. If the limit of the ratio of the derivatives does not exist we cannot conclude nothing.

   Consider as an example:
   $$\lim_{x \to +\infty} \frac{x + \sin x}{x - \cos x} = \frac{\infty}{\infty} = \lim_{x \to +\infty} \frac{x}{x} = 1$$
   We computed the limit applying the elimination principle since $\sin x$ and $\cos x$ are little o of x when $x \longrightarrow +\infty$.

   It is correct to apply the De L'Hopital rule, but it is not helpful:
   $$\lim_{x \to +\infty} \frac{x + \sin x}{x - \cos x} = \lim_{x \to +\infty} \frac{1 + \cos x}{1 + \sin x}$$ and this limit does not exist.

4. Sometimes it is not efficient to apply De L'Hopital because we may easily compute the limit using the significant limits. As an example try to solve the following using both methods:
   $$\lim_{x \to 0} \frac{x^4}{(1 - \cos x)^2}$$

5. In some cases the rule is totally unhelpful:
   $$\lim_{x \to -\infty} \frac{\sqrt{x^2 + 1}}{x} = \frac{\pm \infty}{-\infty} = \lim_{x \to -\infty} \frac{x}{\sqrt{x^2 + 1}} = \lim_{x \to -\infty} \frac{\sqrt{x^2 + 1}}{x}$$
   and as you can realize that we cannot compute the limit using such method.

   One way to solve the problem is the following:
   $$\lim_{x \to -\infty} \frac{\sqrt{x^2 + 1}}{-\sqrt{x^2}} = \lim_{x \to -\infty} -\sqrt{\frac{x^2 + 1}{x^2}} = -1$$

12.6 Differential

Definition 12.6 *A function is said differentiable in c if $\exists k \in \mathbb{R}$:*
$$\lim_{x \to c} \frac{f(x) - [f(c) + k(x - c)]}{x - c} = 0$$

or equivalently:

$$\lim_{\Delta x \to 0} \frac{f(c + \Delta x) - [f(c) + k\Delta x]}{\Delta x} = 0$$

The above definition can be expressed in other words:

a function is differentiable in c if $\exists k \in \mathbb{R}$ when $x \longrightarrow c$:

$$f(x) - [f(c) + k(x - c)] = o(x - c)$$

it means that a function is differentiable if it can be valued using a line with an error going to 0 faster then $(x - c)$; we can also say that the error we observe when approximating $f(x)$ using a line is negligible with rispect to $(x - c)$.

Theorem 12.4 *(The Big Theorem):*

$f(x)$ *is differentiable in* $c \iff f(x)$ *is smooth in* c

(to be differentiable is a sufficient and necessary condition to be smooth).

Proof. $f(x)$ is differentiable in $c \iff$

$$\iff \lim_{x \to c} \frac{f(x) - [f(c) + k(x - c)]}{x - c} = 0 \iff \lim_{x \to c} \left[\frac{f(x) - f(c)}{x - c} - \frac{k(x - c)}{x - c} \right] = 0$$

$$\iff \lim_{x \to c} \frac{f(x) - f(c)}{x - c} = k \iff f'(c) = 0. \blacksquare$$

The proof of the Big Theorem shows us that the slope of the line approximating the function is the first derivative of f and so we are approximating $f(x)$ with a tangent line.

Example 12.9 *Let's compute an approximation for* $\sqrt{101}$.

We consider the function $f(x) = \sqrt{x}$; *its derivative is* $f'(x) = \dfrac{1}{2\sqrt{x}}$.

Now we need a point, close to 101 where we are able to compute $f(x)$ and $f'(x)$; let's take $c = 100$.

Now using the tangent line we have:

$$\sqrt{101} \approx f(c) + f'(c)(x-c) =$$
$$= \sqrt{100} + \frac{1}{2\sqrt{100}}(101 - 100) = 10.05$$

This approximation is quite good; the reader can check, using a calculator that the error is around $0.000\,12$.

12.6.1 Equation of the tangent line

If f is differential in c then the equation of the tangent line to the graph of $f(x)$ in the point c is:

$$y = f(c) + f'(c)(x-c)$$

Then we can write:

$$f(x) - f(c) = f'(c)(x-c) + o(x-c)$$

Let me remind you that

$$m = f'(c) = \tan \alpha$$

that is the slope of the tangent line is equal to the trigonometric tangent of the angle α; α is the angle between the tangent line and the horizontal axis.

Moreover remember that in a right triangle a leg is equal to the other leg times the trigonometric tangent of the opposite angle.

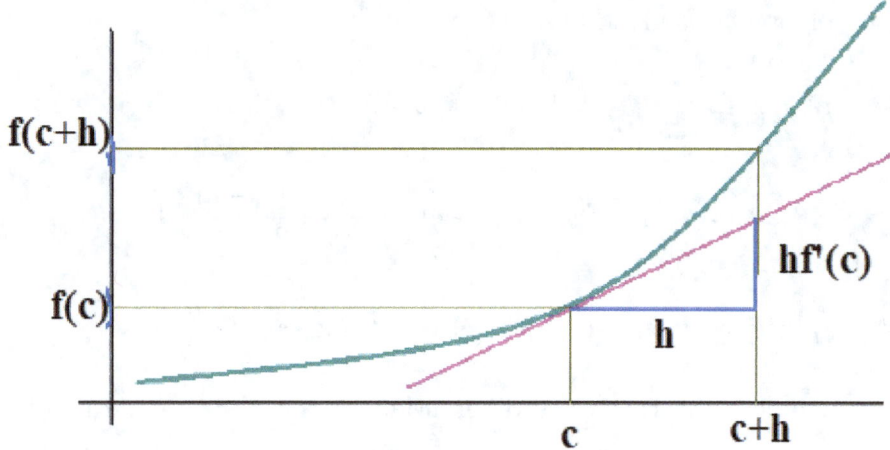

12.7 Series expansion

In the previous section we showed how we can approximate a function using a tangent line (using a first degree polynomial); in this section we'll see how to achieve a better result using a parabola (second degree polynomial) or a higher degree polynomial. The proofs are not presented.

12.7.1 Taylor's polynomial

If the function f can be derived n times in the point c than we can use the Taylor polynomial to approximate the funtion value.

Before presenting the polynomial I am reminding the concept of factorial to the reader:

let $k \in \mathbb{N} \cup \{0\}$ then factorial k is denoted by $k!$ and

$$k! \stackrel{def}{=} \begin{cases} 1 & \text{if } k = 0 \\ k \cdot (k-1)! & \text{if } k > 0 \end{cases}$$

The polynomial is given by the formula:

$$P(n,c) = \sum_{k=0}^{n} \frac{f^{(k)}(c)(x-c)^k}{k!} =$$

$$= f(c) + f^{(1)}(c)(x-c) + \frac{f^{(2)}(c)(x-c)^2}{2} + ...$$

$f^{(k)}$ is f derived k times; so $f^{(0)}(c)$ is nothing but $f(c)$.

The first degree term is the differential; so the first two terms of the polynomial represent the equation of the tangent line to the graph of the function. The first three terms are a second degree polynomial wich represent the tangent parabola to the graph and so on.

Using such a polynomial we can value a function in a neighborhood of c.

Example 12.10 *Let's compute a rounded value for* $\log 2$ *using a Taylor polynomial of degree 5.*

Now we use the function $f(x) = \log(1+x)$; we choose the point $c = 0$. So we have:

k	$f^{(k)}(x)$	$f^{(k)}(c)$	$k!$
0	$\log(1+x)$	0	1
1	$\frac{1}{1+x}$	1	1
2	$-\frac{1}{(1+x)^2}$	1	2
3	$\frac{2}{(1+x)^3}$	1	6
4	$-\frac{6}{(1+x)^4}$	1	24
5	$\frac{24}{(1+x)^5}$	1	120

As usual I strongly suggest to compute these derivatives.

The polynomial is

$$P(5,1) = (x-1) - \frac{1}{2}(x-1)^2 + \frac{1}{3}(x-1)^3 ...$$

and the rounded value is

$$\log 2 \approx 1 - \frac{1}{2} + \frac{1}{3} - \frac{1}{4} + \frac{1}{5} = 0.78\overline{3}$$

We consider the reminder (error) of the Taylor'polynomial the difference:

$$R(n, x - c) = f(x) - P(n, c)$$

Of course the reminder depends on the distance between x and c (it is small if x is close to c) and on the degree of the polynomial (under certain conditions it becomes smaller and smaller as the degree increases).

We now consider two particular formulations of the reminder due to the Italian mathematicians Lagrange and Peano.

Lagrange's reminder:

Lagrange's reminder is a quantitative estimate of the error that occurs when approximating the function value using a polynomial. We have:

$$R(n, x - c) = \frac{1}{(n+1)!} f^{(n+1)}(\gamma)(x - c)^{n+1}$$

with $\gamma \in \,]c, x[$.

Example 12.11 *Let's refer to example 12.10; we have*

$$R(6, x - 1) = \frac{1}{6!} \left[-\frac{120}{(1+\gamma)^6} \right] (x-1)^6 = -\frac{1}{6} \frac{(x-1)^6}{(1+\gamma)^6}$$

We don't know γ, but in the specific example it is between 0 and 1; it follows that

$$|R(6, 2-1)| < \frac{1}{6} = 0.1\overline{6}$$

The upper extimate of the error has been computed considering $\gamma = 0$.

Indeed a better approximation of the error has been computed using a computer software and is equal to -0.09019 (much smaller in absolute value than the maximum estimate obtained by Lagrange's reminder).

Peano's Reminder

Peano's Reminder is a qualitative extimate of the error; it is

$$R(n, x - c) = o\left[(x - c)^n\right]$$

Example 12.12 *Let's compute the following limit*

$$\lim_{x \to 0} \frac{x - \sin x}{x^3} = \frac{0}{0}$$

It is not possible to solve the limit using significant limits, so let's try using the Taylor's polynomial centered in 0.

Since the biggest exponent for the x is 3 we can use a polynomial of third degree for the function $\sin x$ since what remains is negligible; since the x tends to 0 we use $c = 0$.

We have

k	$f^{(k)}(x)$	$f^{(k)}(c)$	$k!$
0	$\sin x$	0	1
1	$\cos x$	1	1
2	$-\sin x$	0	2
3	$-\cos x$	-1	6

and it follows

$$\lim_{x \to 0} \frac{x - \sin x}{x^3} = \lim_{x \to 0} \frac{x - \left[x - \frac{1}{6}x^3 + o(x^3)\right]}{x^3} = \frac{1}{6}$$

12.7.2 McLaurin's polynomial

McLaurin's polynomial is not so different from Taylor's polynomial: the only difference is that now $c = 0$. So we can write:

$$P(n,0) = \sum_{k=0}^{n} \frac{f^{(k)}(0) x^k}{k!} =$$
$$= f(0) + f^{(k)}(0) x + \frac{f^{(2)}(0) x^2}{2} + \frac{f^{(3)}(0) x^3}{2} + ...$$

Of course we can still apply both reminder definitions we used about Taylor's polynomial.

Exercise 12.3 *Write the Mc Laurin polynomial of 5th degree for the function e^x; use the polynomial to get an approximation of e^2.*

12.8 Rules to determine greatest and smallest of a function

$1^{S(t)}$ **rule** if in the point c the function is continuous and its derivative changes the sign then c is an extreme (a local maximum or minimum). If the derivative is negative before c and positive after c the point is a minimum; if the derivative is positive before c and negative after c the point is a maximum.

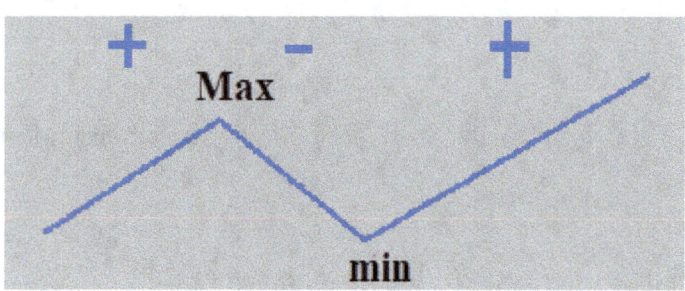

2nd rule if in the point c the function is continuous, the first derivative is equal to zero and the derivative of the second order is positive (negative) then c is a local minimum (maximum) point.

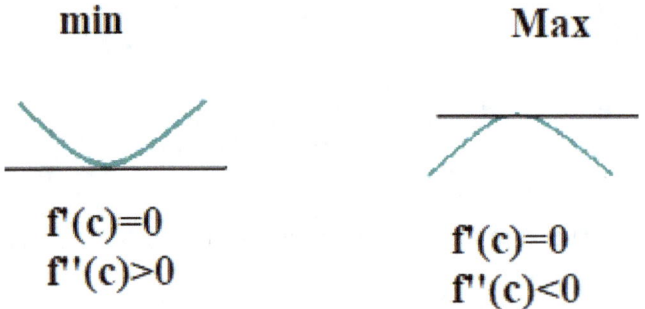

The proof of the first rule follows from simple geometric considerations, while to prove the second one we can use Taylor's polynomial (of the second degree) in the point c using Peano's reminder; if the first derivative is equal to 0 in c we have:

$$f(x) = f(c) + f''(c)(x-c)^2 + o\left((x-c)^2\right)$$

and since the reminder is negligible with respect to the term of the 2nd degree we can write:

$$f(x) \approx f(c) + f''(c)(x-c)^2$$

hence if $f''(c)$ is positive it follows $f(x) > f(c)$ in a neighborhood of c, and c is a local minimum point; if $f''(c)$ is negative then $f(x) < f(c)$ in a neighborhood of c, and it is a local maximum point.

We cannot conclude nothing if the second derivative is equal to 0.

12.9 Concave an convex functions

A set is convex if the segment joining any couple of its points is fully contained in the set.

convex set   non convex set

If the set is a subset of $\mathbb{R}$ the segment joining the points a and b is:

$$\alpha a + (1-\alpha) b$$

with $\alpha \in [0;1]$.

Convex subset of $\mathbb{R}$ are points, interval (both open and closed) the half-lines and $\mathbb{R}$. If the set is a subset of $\mathbb{R}^2$ the segment joining two points whoose coordinates are $(x_a; y_a)$ and $(x_b; y_b)$ is:

$$(\alpha x_a + (1-\alpha)x_b; \alpha y_a + (1-\alpha)y_b)$$

with $\alpha \in [0;1]$

The upper graph of $f(x)$ is the set of points $(x;y)$ such that $y \geq f(x)$.

The lower graph of $f(x)$ is the set of points $(x;y)$ such that $y \leq f(x)$.

A function is said convex in A (subset of its domain) if the upper graph, restricted to A, is a convex set.

A function is said concave in A (subset of its domain) if the lower graph, restricted to A, is a convex set.

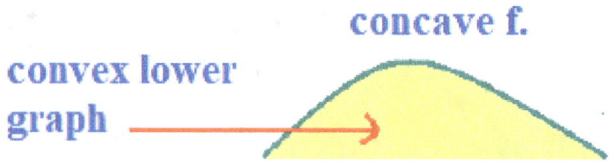

We can also say that f is convex in A if

$$f(\alpha x_1 + (1-\alpha)x_2) \leq \alpha f(x_1) + (1-\alpha)f(x_2)$$

with $\alpha \in [0;1]$, x_1 and $x_2 \in A$.

f is concave in A if:

$$f(\alpha x_1 + (1-\alpha)x_2) \geq \alpha f(x_1) + (1-\alpha)f(x_2)$$

with $\alpha \in [0;1]$, x_1 and $x_2 \in A$.

A differentiable function in c is convex in c if its graph is above its tangent line:

$$\exists N(c) : f(x) \geq f(c) + f'(c)(x-c) \, \forall x \in N(c)$$

The differentiable function f is concave in c if the graph is below the tangent line:

$$\exists N(c) : f(x) \leq f(c) + f'(c)(x-c) \, \forall x \in N(c)$$

Using Taylor's polynomial we can see that if we can derive $f(x)$ twice and if the derivative of the second order is positive then it is convex; if $f''(x) < 0$ then it is concave.

Remark 12.2 *1. increasing convex functions increase faster and faster;*

2. decreasing convex functions decrease slower and slower;

3. concave increasing functions increase slower and slower;

4. convex decreasing functions decrease faster and faster.

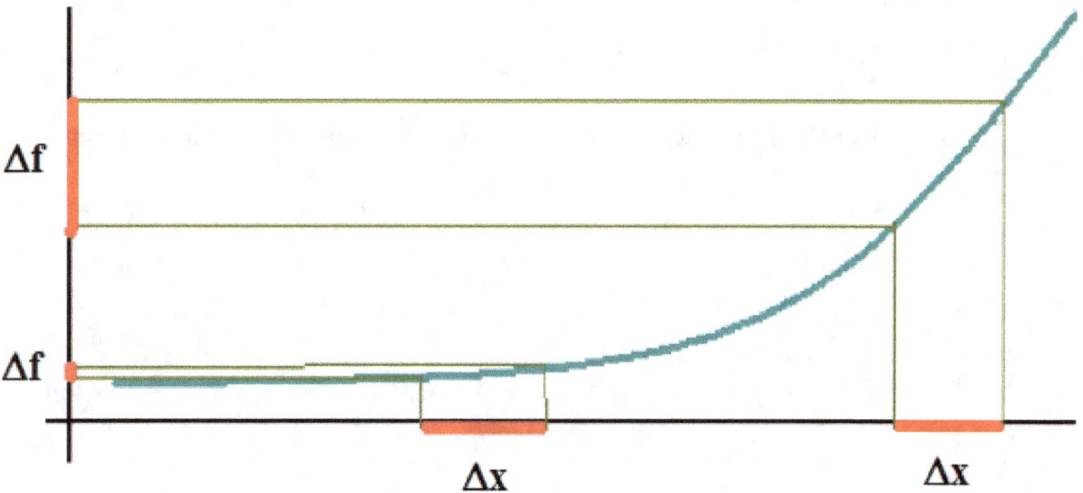

From the picture we can see that given an increasing and convex function the same variation in the independent variable (Δx) determines a small increment at the beginning, a larger increment later.

Remark 12.3 *Referring to the sorting of the infinite (page 137) please note that when $x \longrightarrow +\infty$ function that are infinite of greater order than x are convex while function that are infinite of lower order than x are concave.*

Now we can easily understand that:

1. [f is increasing and convex] $\implies$ [f increases faster and faster] $\implies$

    $\implies$ [the slope of the tangent line is increasing] $\implies$ [f' is increasing] $\implies$

    $\implies [f'' > 0]$.

2. [f is decreasing and convex] $\implies$ [f decreases slower and slower] $\implies$

    $\implies$ [the slope of the tangent line is increasing] $\implies$ [f' is increasing] $\implies$

    $\implies [f'' > 0]$.

3. [f is increasing and concave] $\implies$ [f increases slower and slower] $\implies$

    $\implies$ [the slope of the tangent line is decreasing] $\implies$ [f' is decreasing] $\implies$

    $\implies [f'' < 0]$

4. [f decreasing and concave] $\implies$ [f decreases faster and faster] $\implies$

    $\implies$ [the slope of the tangent line is decreasing] $\implies$ [f' is decreasing] $\implies$

    $\implies [f'' < 0]$

Chapter 13

Graph of a function

13.1 Steps to plot a function

To plot the graph of a function we should execute (if possible) the following steps:

1. First of all we have determine the domain (usually I provide this information to my students); some fundamental rules are:

    - $\sqrt[2n]{g(x)}$ exists if and only if $g(x) \geq 0$;
    - $\log[g(x)]$ exists if and only if $g(x) > 0$;
    - $\dfrac{a}{g(x)}$ exists if and only if $g(x) \neq 0$.

2. Intersections with the cartesian axes (if they exist and if we are able to determine them); we compute them as follows:

    - vertical axis: we compute $f(0)$ (if 0 belongs to the domain);
    - horizontal axis: we solve the equation $f(x) = 0$.

3. Sign of the function; we solve $f(x) > 0$, this point can be treated together with the previous one solving $f(x) \geq 0$;

4. Limits in the limit (accumulation) points of the domain that don't belong to the domain;

5. Compute the first derivative to determine where the function is inreasing or decreasing, the extrema of the function (minima and maxima), turning point with horizontal tangent line (stationary points);

6. Identification of the points where the function is continuous but not smooth;

7. Determine where the function is convex or concave; this can be done studying the sign of the second derivative or analyzing the magnitude of the function and other features of the function (much better in my opinion).

13.2 Solved exercises

13.2.1 $f(x) = \frac{\log x + 1}{\log x - 1}$

The domain is $\mathbb{R}_+ - \{0, e\}$ because $[x > 0]$ and $[(\log x - 1) \neq 0]$ must hold; the function never crosses the vertical axis.

$\frac{\log x + 1}{\log x - 1} \geq 0$ when:

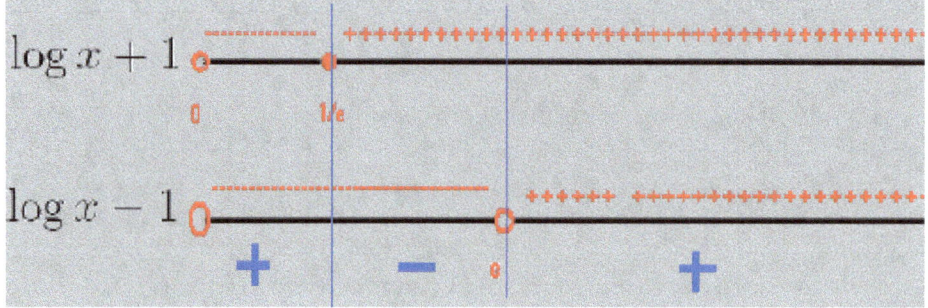

$$\lim_{x \to +\infty} \frac{\log x + 1}{\log x - 1} = \frac{+\infty}{+\infty} = \lim_{x \to +\infty} \frac{\log x}{\log x} = 1 \text{ and the line } y = 1 \text{ is horizontal asymptote;}$$

$$\lim_{x \to 0^+} \frac{\log x + 1}{\log x - 1} = \frac{-\infty}{-\infty} = \lim_{x \to 0^+} \frac{\log x}{\log x} = 1$$

(please notice the application of the elimination principle).

$$\left. \begin{array}{l} \lim_{x \to e^+} f(x) = \frac{2}{0^+} = +\infty \\ \lim_{x \to e^-} f(x) = \frac{2}{0^-} = -\infty \end{array} \right\} \text{ hence the line } x = e \text{ is vertical asymptote;}$$

moreover there exists a right neighborhood of e where the function is convex (landing from $+\infty$ at the beginning it must decrease very very fast), there exists a left neighborhood of e where the function is concave (it must decrease very very fast to go down to $-\infty$).

$f'(x) = -\frac{2}{x(\log x - 1)^2} < 0 \forall x \in D_f$ and the function is decreasing everywhere; in a neighborhood of $+\infty$ the function is convex since it must decrease slower and slower approaching the asymptote.

$\lim_{x \to 0^+} f'(x) = -\infty$ hence the tangent line to the graph becomes vertical and in a right neighborhood of zero the function is convex, there is at least a turning point in $]0; e[$.

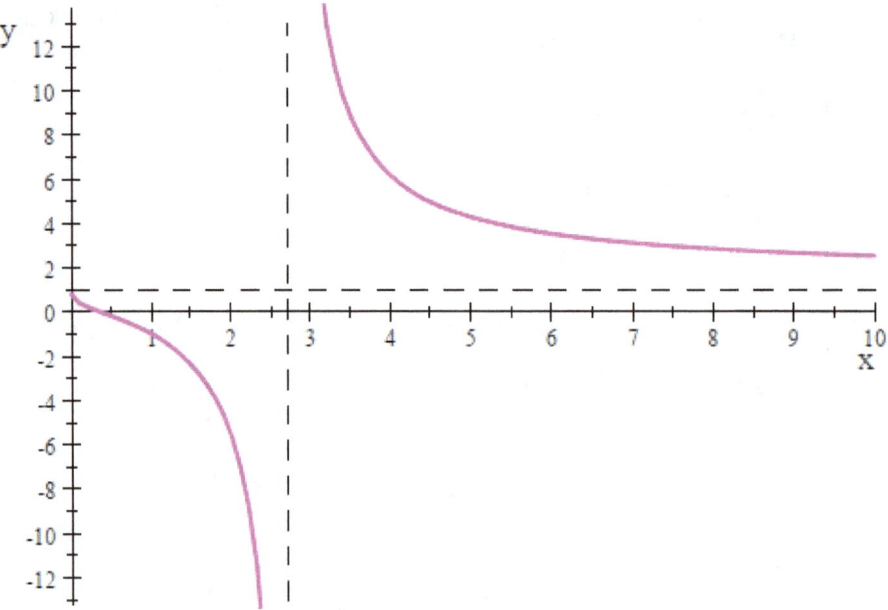

The second derivative is not too difficult and prove that there is a unique turning point corresponding to the intersection with the horizontal axis.

13.2.2 $f(x) = xe^x$

$\mathbb{R}$ is the domain, $(0;0)$ belongs to the graph of the function, f is negative when $x < 0$ and is positive when $x > 0$;

$\lim\limits_{x \longrightarrow -\infty} xe^x = -\infty \, (0^+) = 0$ and so the line $y = 0$ is horizontal asypmtote;

$\lim\limits_{x \longrightarrow +\infty} xe^x = +\infty$.

$f'(x) = e^x(1+x) > 0$ when $x > -1$ and so -1 is a global minimum point;

$f(-1) = -1/e$ (minimum of the function).

The function is smooth in the minimum point; hence it is convex in a neighborhood of -1; it is convex in a neighborhood of $+\infty$ since it increases very fast ($e^x = o(f(x))$)

and in a neighborhood of $-\infty$ it is concave because of the horizontal asymptote: at the beginning the function decreases very very slowly. There is at least a turning point in $]-\infty; -1[$.

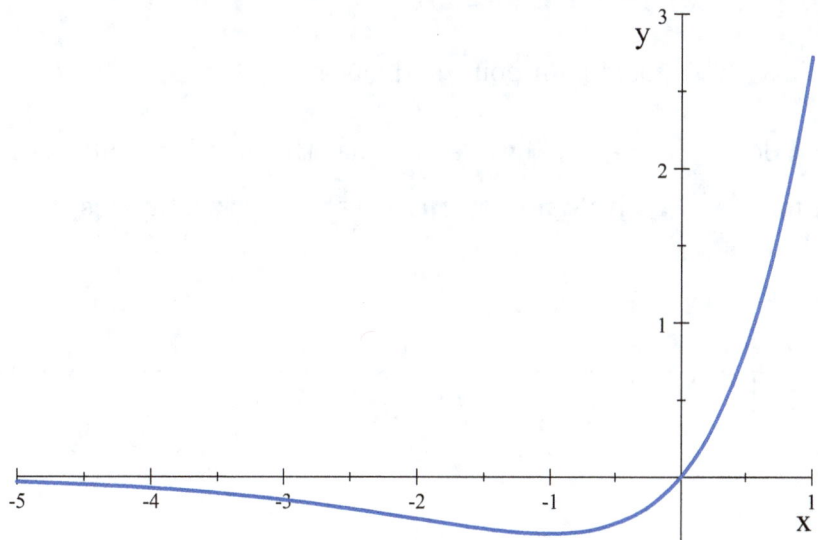

The second derivative is very easy and it shows that there is a turning (inflection) point in -2.

13.3 $f(x) = \log x - \frac{1}{2}x^2$

The domain is given by the condition $x > 0$; I'm not able to solve neither $f(x) = 0$ nor $f(x) > 0$....

$$\lim_{x \to 0^+} \left[\log x - \tfrac{1}{2}x^2\right] = -\infty + 0 = -\infty$$

$$\lim_{x \to +\infty} \left[\log x - \tfrac{1}{2}x^2\right] = +\infty - \infty = -\infty$$

since $\log x = o\left(\tfrac{1}{2}x^2\right)$.

197

From these limits we can realize (applying the sign theorem) that the function is negative in a right neighborhood of 0 and in a neighborhood of $+\infty$.

The line $x = 0$ is a vertical asymptote.

$f'(x) = \frac{1-x^2}{x} \geq 0$ when $0 < x \leq 1$;

so 1 is a global maximum point and since:

$f(1) = \log 1 - \frac{1}{2} = -\frac{1}{2} < 0$ we realize that the function is always negative (the global maximum is negative) and never crosses the horizontal axis.

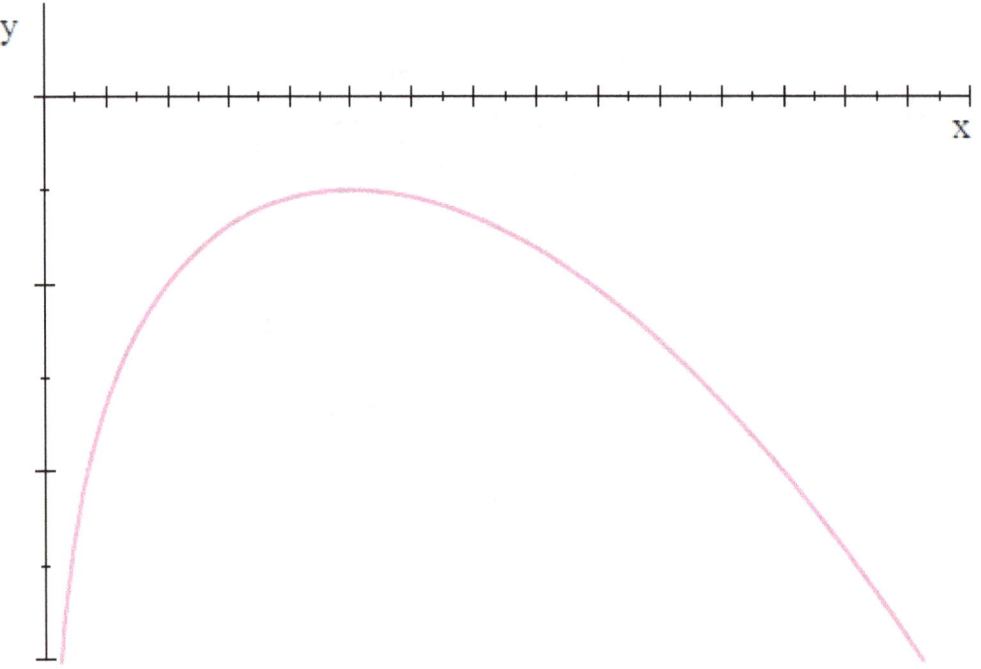

Since $x = 0$ is vertical asymptote we know that in a right neighborhood of zero the function is concave (at the beginning it grows very very fast), concave in a neighborhood of the maximum point and concave in a neighborhood of $+\infty$ since the function goes to $-\infty$ but it is an infinite of bigger magnitude than x. These considerations make me think that the function has no turning points; anyway the second derivative is very easy:

$f''(x) = -\dfrac{1}{x^2}$ and is always negative; hence the function is concave everywhere in the domain.

13.4 $f(x) = e^x \cdot \sqrt[3]{x^2}$

The domain is the set of the real numbers; the function is positive everywhere and $f(0) = 0$; hence 0 is a minimum point and the global minimum of the function is 0.

$$\lim_{x \to -\infty} e^x \sqrt[3]{x^2} = 0$$

$$\lim_{x \to +\infty} e^x \sqrt[3]{x^2} = +\infty$$

The line $y = 0$ is horizontal asymptote.

$$f'(x) = \dfrac{e^x (3x + 2)}{3\sqrt[3]{x}}$$

Sign of the first derivative:

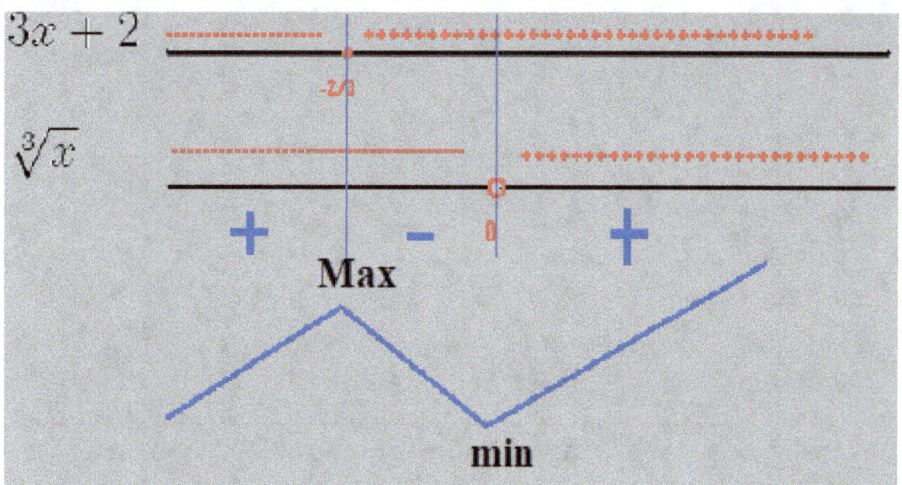

Please pay attention to what follows: in the maximum point the derivative is zero (horizontal tangent line); in the minimum point the derivate does not exist because the denominator is zero. To understand the behavior of the function in the point where it is not smooth we can compute the following limits:

$$\lim_{x \to 0^-} f'(x) = -\infty$$

$$\lim_{x \to 0^+} f'(x) = +\infty$$

So in 0 there is a cusp; the graph, in a neighborhood of the origin is something like:

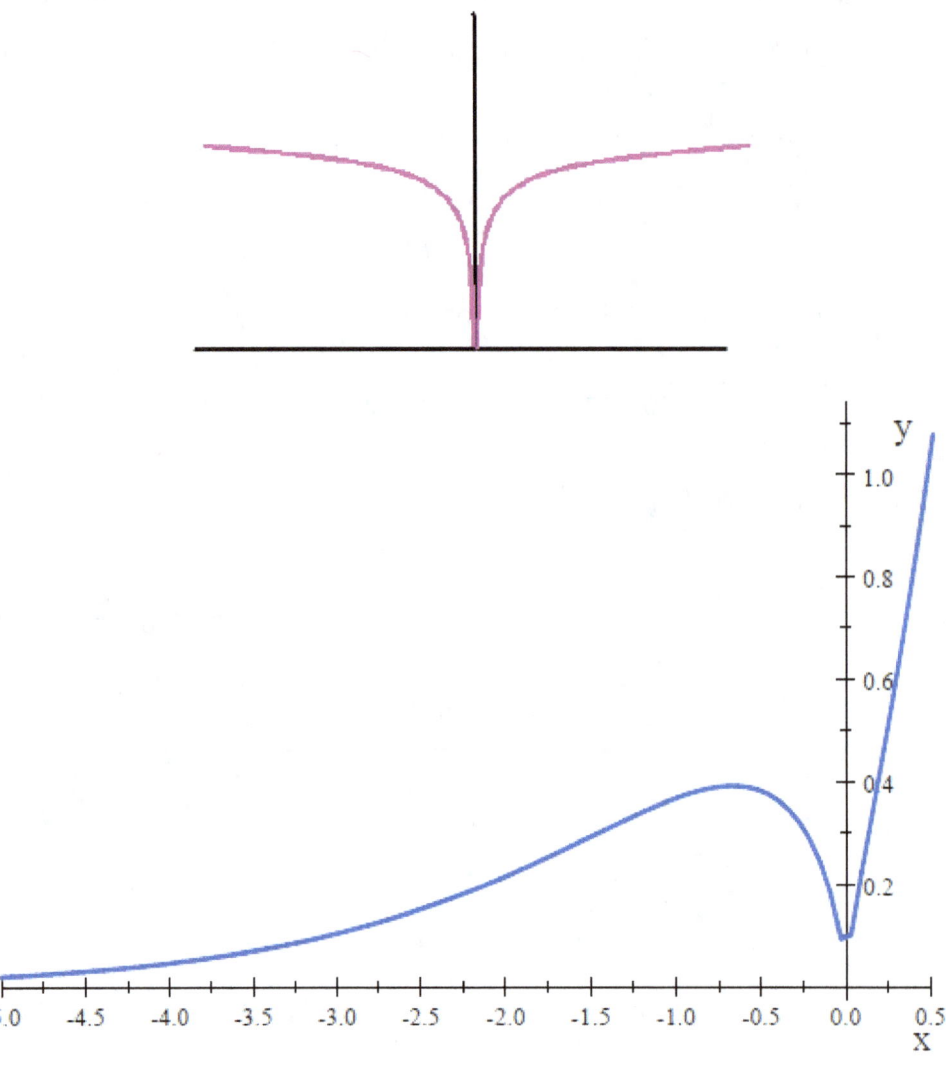

In a neighborhood of $+\infty$ the function is convex: $e^x = o(f(x))$ when $x \longrightarrow +\infty$ hence the function is increasing faster and faster: increasing convex functions has this property. It follow that a turning point must exist when $x > 0$, to know its value we should compute f'' (not an impossible mission but not so easy). In a neighborhood of -1 the function is convex (because of the horizontal asymptote...). In the maximum point the function is smooth, and so in a neighborhood of $-2/3$ the function must be concave; as a consequence at least a turning point is in $]-\infty; -2/3[$.

We cannot conclude anything else without computing the second derivative.

13.5 $f(x) = \dfrac{\log^2 x}{x^3}$

The domain is the set of the strictly positive real numbers (input of the logarithm strictly greater than zero), the function is always positive and $f(1) = 0$; it means that 1 is a global minimum point.

$$\lim_{x \to +\infty} \frac{\log^2 x}{x^3} = \frac{+\infty}{+\infty} = 0$$

since the infinite in the denominator is of bigger magnitude with respect to the numerator and $y = 0$ is horizontal asymptote;

$$\lim_{x \to 0^+} \frac{\log^2 x}{x^3} = \frac{+\infty}{0^+} = +\infty$$

and the line $x = 0$ is vertical asymptote.

$$f'(x) = \frac{2 \log x \cdot \frac{1}{x} x^3 - \log^2 x \cdot 3x^2}{x^6} = \frac{\log x \, (2 - 3 \log x)}{x^4}$$

$f'(x) \geq 0$ when $\log x \, (2 - 3 \log x) \geq 0$:

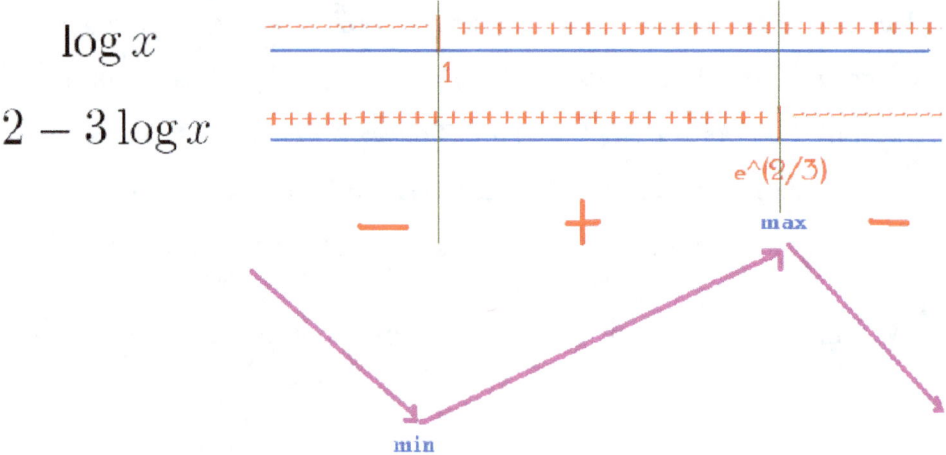

In the maximum point and in the minimum one the derivative is zero (horizontal tangent line); the function is smooth in the stationary points hence it is concave in a neighborhood of the maximum point and convex in a neighborhood of the minimum point; the asymptotes help us to understand that the function is convex in a right neighborhood of 0 and in a neighborhood of $+\infty$:

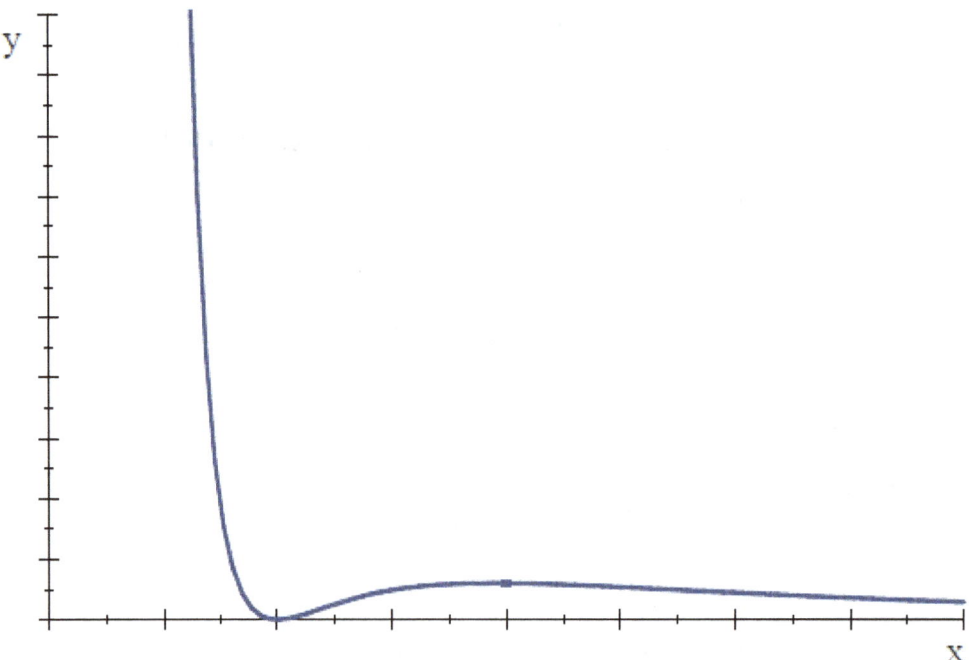

We have at least two turning points: one is between minimum and maximum points, the other one somewhere on the right of the maximum point.

It is hard but not impossible to compute the second derivative; it is:

$$f''(x) = \frac{6\log^2 x - 7\log x + 1}{x^5}$$

The denominator is always positive in the domain; we can solve $6\log^2 x - 7\log x + 1 \geq 0$ setting $\log x = t$:

$$6t^2 - 7t + 1 \geq 0$$

we can solve the equation $6t^2 - 7t + 1 = 0$, solutions are: 1 and $\frac{1}{6}$; the inequality is satisfied when:

$$t \leq \frac{1}{6} \iff \log x \leq \frac{1}{6} \iff 0 < x \leq e^{\frac{1}{6}} = \sqrt[6]{e}$$

or

$$t \geq 1 \iff \log x \geq 1 \iff x > e$$

the function is convex where f'' is positive; e and $\sqrt[6]{e}$ are turning points.

13.6 $f(x) = e^{\frac{1}{x^2-1}}$

Domain:

$$[x^2 - 1 \neq 0] \iff x \neq \pm 1$$

The function is always positive and since $f(0) = e^{-1}$ the graph crosses the vertical axis in $(0; 1/e)$;

$$\lim_{x \to +\infty} f(x) = \lim_{x \to -\infty} f(x) = e^{\frac{1}{+\infty}} = e^0 = 1$$

The line $y = 1$ is horizontal asymptote;

$$\lim_{x \to 1^+} f(x) = \lim_{x \to -1^-} f(x) = e^{\frac{1}{1^+ - 1}} = e^{\frac{1}{0^+}} = e^{+\infty} = +\infty$$

The lines $x = 1$ and $x = -1$ are vertical asymptotes;

$$\lim_{x \to 1^-} f(x) = \lim_{x \to -1^+} f(x) = e^{\frac{1}{1^- - 1}} = e^{\frac{1}{0^-}} = e^{-\infty} = 0$$

$$f'(x) = e^{\frac{1}{x^2-1}} \cdot \frac{-2x}{(x^2-1)^2}$$

it is positive iff the x is negative hence the function is increasing when $x < 0$ and in zero there is a local maximum point.

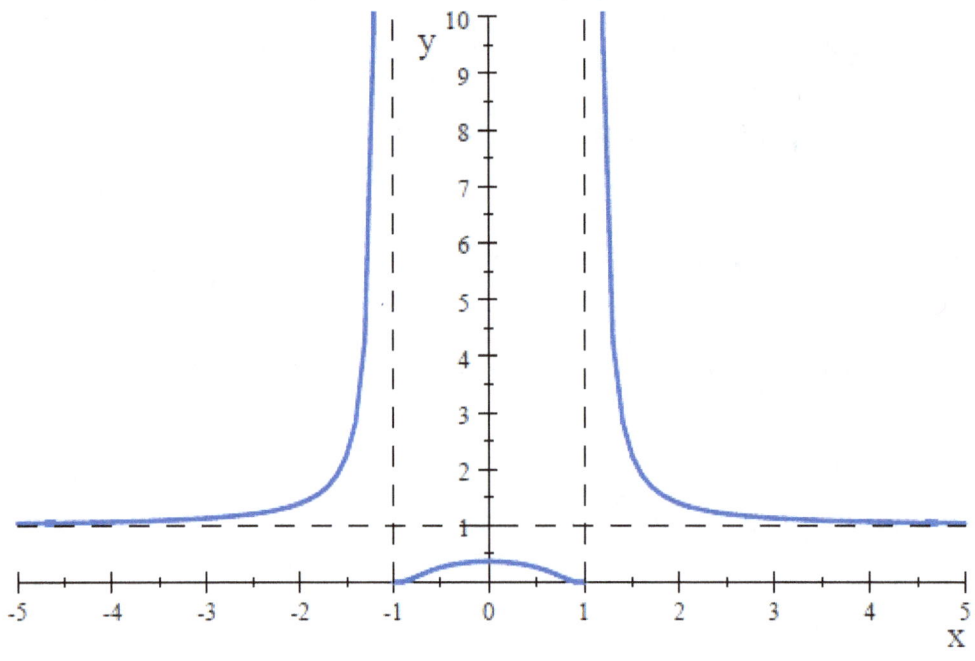

13.7 $f(x) = \frac{x}{\log x}$

The domain is the set of the x greater than zero and different by 1 ($\log x \neq 0$); the function is positive when $x > 1$.

We have:

$$\lim_{x \to 1^+} \frac{x}{\log x} = \frac{1}{0^+} = +\infty$$

$$\lim_{x \to 1^-} \frac{x}{\log x} = \frac{1}{0^-} = -\infty$$

The line $x = 1$ is vertical asymptote.

$$\lim_{x \to 0^+} \frac{x}{\log x} = \frac{0^+}{-\infty} = 0$$

$$\lim_{x \to +\infty} \frac{x}{\log x} = \frac{+\infty}{+\infty} = +\infty$$

since the numerator is an infinite of bigger magnitude than the denominator.

$$f'(x) = \frac{\log x - 1}{\log^2 x}$$

$$[f'(x) \geq 0] \iff [\log x - 1 \geq 0] \iff [x \geq e]$$

hence:

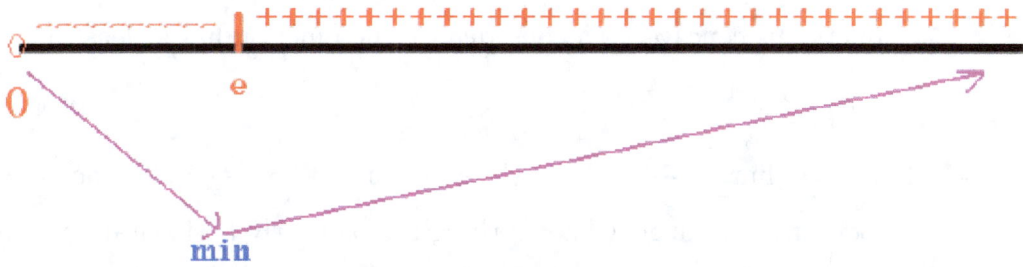

and $f(e) = e$;

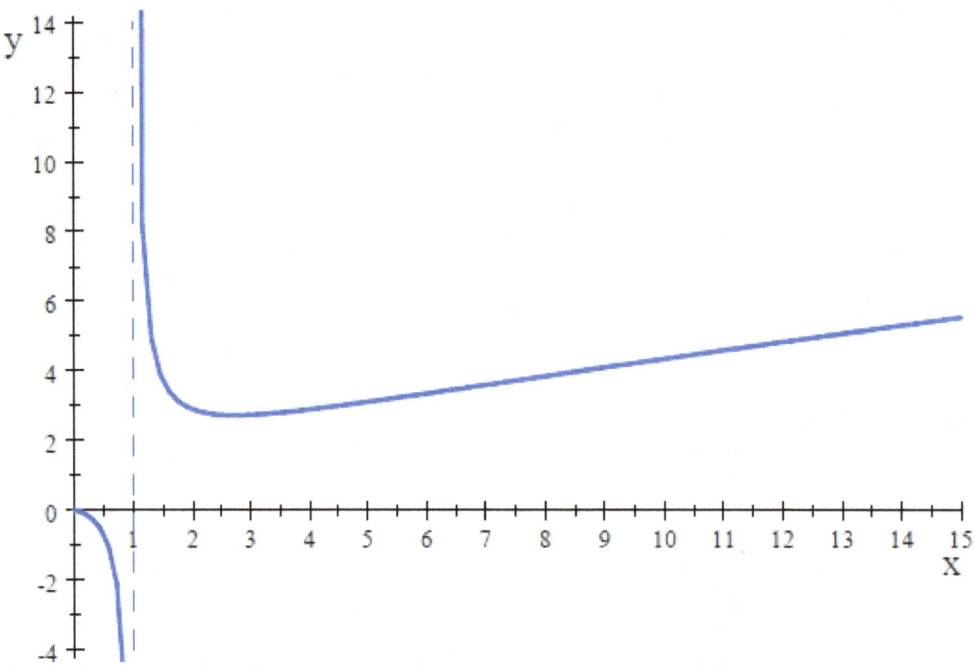

To understand where the function is convex or concave we have to consider:

- in a neighborhood of the minimum point the (smooth) function is convex;

- because of the vertical asymptote there exists a left neighborhood of 1 where the function is concave and a right neighborhood of 1 where la function is convex;

- $f(x) = o(x)$ when $x \longrightarrow +\infty$ hence there exists a neighborhood of $+\infty$ where the function is concave; as a consequence the function has at least a turning point in $]2; +\infty[$;

- $\lim\limits_{x \longrightarrow 0+} y' = \lim\limits_{x \longrightarrow 0+} \frac{\log x - 1}{\log^2 x} = \lim\limits_{x \longrightarrow 0+} \frac{\log x}{\log^2 x} = \lim\limits_{x \longrightarrow 0+} \frac{1}{\log x} = \frac{1}{-\infty} = 0$ hence in a neighborhood of 0 the tangent line to the graph is nearly horizontal and exists a right neighborhood of 0 where the function is concave.

Chapter 14

Integrals

In these chapter we face the problem of determining the area of a region underlying the graph of a function. This is the main concept the reader should keep in mind reading this chapter and, later, facing applications.

14.1 Reinann's definition of integral

Let $f(x)$ be a bounded function defined in the interval $[a,b]$, and let $\mathcal{L}$ be a partition set of $[a;b]$:

$$\mathcal{L} := \{x_0, x_1, ..., x_n\}$$

where: $x_0 = a,\quad x_n = b,\quad x_{k-1} < x_k$.

$\mathcal{L}$ identifies the sub-intervals of $[a,b]$: $[x_0, x_1], [x_1, x_2], ..., [x_{n-1}, x_n]$

Definition 14.1 *Lower sum depending on $\mathcal{L}$:*

$$S_\mathcal{L}^i = \sum_{k=1}^n (x_k - x_{k-1}) \cdot f_k^i$$

where f_k^i is the infimum of f in the kth sub-interval. Consider the following picture:

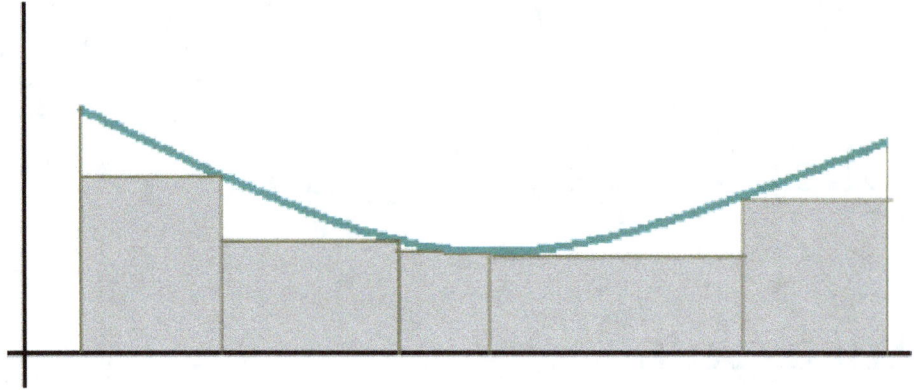

Definition 14.2 *Upper sum depending on $\mathcal{L}$:*

$$S_\mathcal{L}^s = \sum_{k=1}^n (x_k - x_{k-1}) \cdot f_k^s$$

where f_k^s is the supremum of f in the kth sub-interval. Look at the picture:

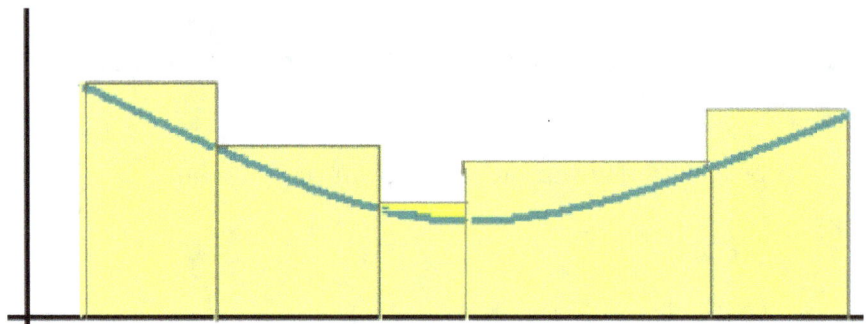

We can also say that the lower sum is the area of the inner step region and the upper sum is the area of the outer step region.

Proposition 14.1

$$\forall \mathcal{L} \ S_\mathcal{L}^i \leq S_\mathcal{L}^s$$

That is the lower sum depending on $\mathcal{L}$ is smaller than the corresponding upper sum (depending on the same partition). The proof is very easy and is based on the inequalities $f_k^i \leq f_k^s \forall k$.

Definition 14.3 *The partition $\mathcal{D}$ is said thinner than $\mathcal{L}$ if $\mathcal{L} \subseteq \mathcal{D}$ (we can also say that $\mathcal{D}$ is a refinement of $\mathcal{L}$).*

Proposition 14.2

$$[\mathcal{L} \subseteq \mathcal{D}] \implies \begin{cases} S_\mathcal{L}^i \leq S_\mathcal{D}^i \\ S_\mathcal{L}^s \geq S_\mathcal{D}^s \end{cases}$$

that is: when the partition becomes thinner the lower sums increase while the upper sums decrease (let me remind you that $\subseteq$ is a partial weak order and so we can say that lower sums increase when the partition increases).

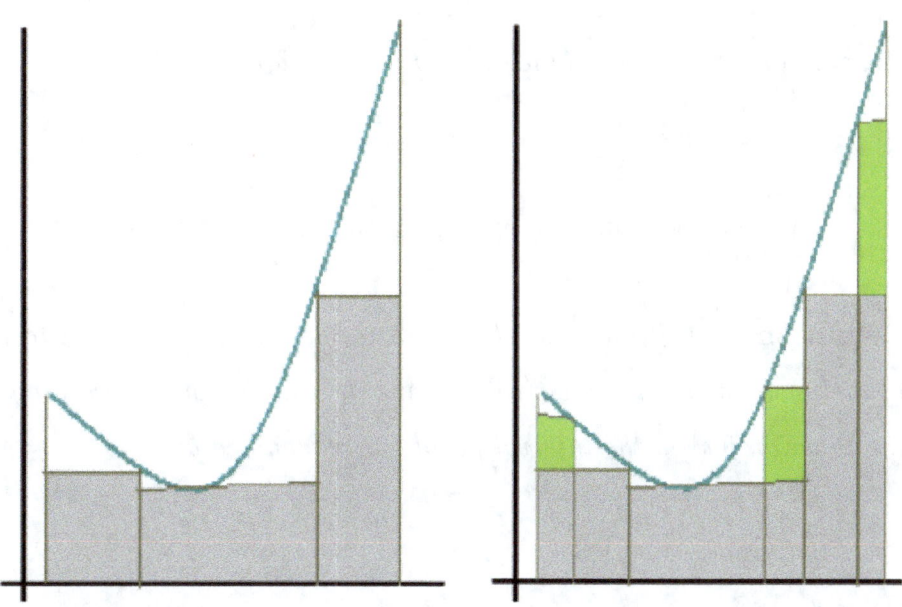

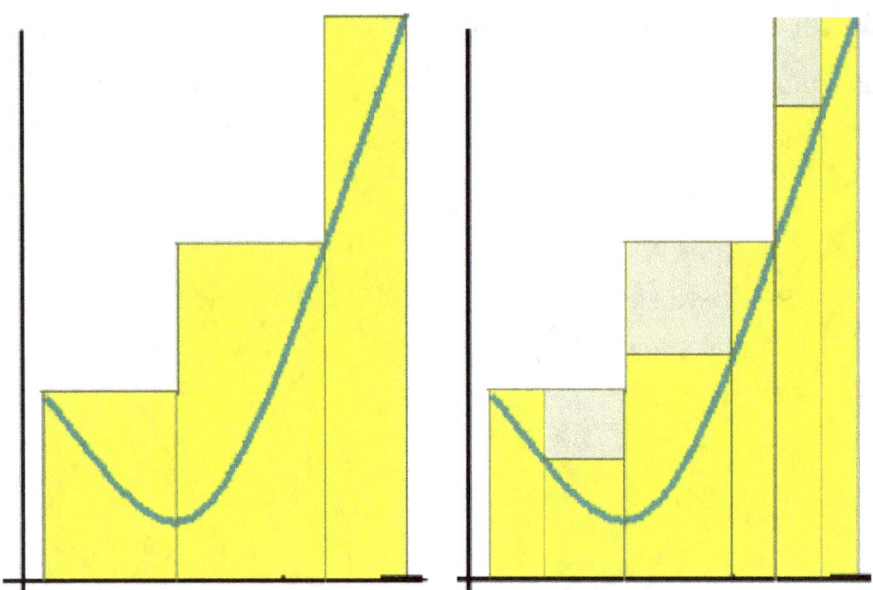

Proposition 14.3

$\forall \mathcal{L}, \mathcal{D}$ we have $S^i_{\mathcal{L}} \leq S^s_{\mathcal{D}}$

that is any upper sum is an upper bound for the set of the lower sums and any lower sum is a lower bound of the set of the upper sums.

Proof. since $\mathcal{L} \subseteq (\mathcal{L} \cup \mathcal{D})$ and $\mathcal{D} \subseteq (\mathcal{L} \cup \mathcal{D})$ it follows:

$$S^i_{\mathcal{L}} \leq S^s_{\mathcal{L}} \leq S^s_{\mathcal{L} \cup \mathcal{D}} \leq S^s_{\mathcal{D}}$$

and because of the transitive properties the proof is complete. ∎

Definition 14.4 *if the upper bound of the lower sums is equal to the lower bound of the upper sums then their value is the definite integral (according to Reimann) of the function f(x) in the interval [a; b] and is denoted as:*

$$\int_a^b f(x)\,dx$$

Remark 14.1 *Please note that there are some similarities between the notations $\sum_{k=1}^{n}(x_k - x_{k-1}) \cdot f_k$ and $\int_a^b f(x)\, dx$:*

- the symbol $\int$ is a stylized s and $\sum$ is the Greek letter for S (capital case) representing the sum;

- a and b are called integration bounds (a is the lower bound and b is the upper one) and take the place of 1 and n meaning that the sum has to be taken all over $[a, b]$;

- $f(x)$ is the value of the function and replaces f_k;

- dx is the infinitesimal width of each interval and is replacing $(x_k - x_{k-1})$.

14.2 Properties of the integral

1. $\int_a^b [f(x) + g(x)]\, dx = \int_a^b f(x)\, dx + \int_a^b g(x)\, dx$

2. $\int_a^b k f(x)\, dx = k \int_a^b f(x)\, dx$

3. $\int_a^b f(x)\, dx = -\int_b^a f(x)\, dx$ (by definition)

4. $\int_a^a f(x)\, dx = 0$

5. $\int_a^b f(x)\, dx = \int_a^c f(x)\, dx + \int_c^b f(x)\, dx$

    The equality holds both if c in the interior of $[a; b]$ and if c is outside the interval. In fact:

- case $a < c < b$:

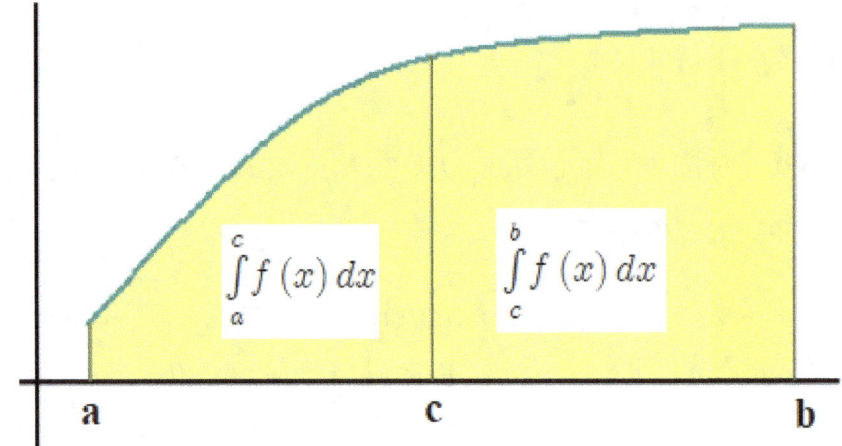

- case $a < b < c$:

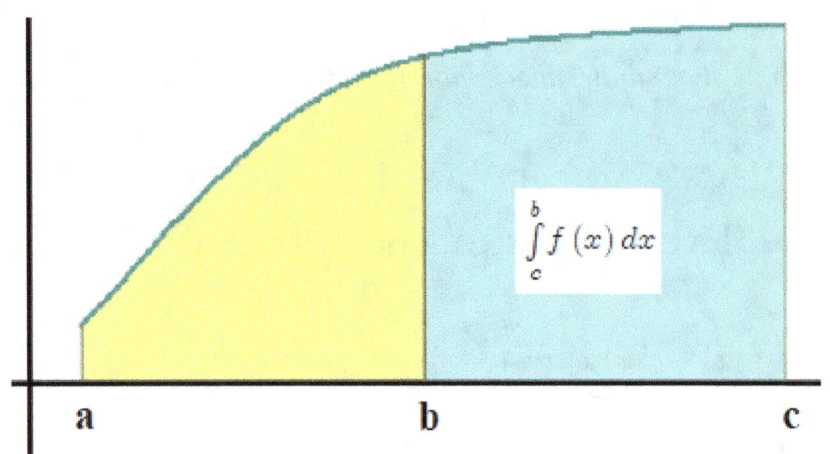

Applying property number 3 we can write:

$$\int_a^b f(x)\,dx = \int_a^c f(x)\,dx - \int_b^c f(x)\,dx = \int_a^c f(x)\,dx + \int_c^b f(x)\,dx$$

14.3 Theorems on integrals

Here we present three sufficient conditions for the existence of the definite integral (the proofs are not presented):

1. $f(x)$ is continuous in a closed bounded interval;

2. $f(x)$ is bounded with a finite number of discontinuities;

3. $f(x)$ is bounded with a (infinite) countable number of discontinuities.

Theorem 14.1 *Mean value theorem (by Lagrange)*

$$f(x) \text{ is continuous in } [a;b]$$
$$\Downarrow$$
$$\exists c \in \,]a;b[\,: \int_a^b f(x)\,dx = (b-a)\,f(c)$$

Proof. if $f(x)$ is continuous in $[a;b] \implies$ (Weierstrass's Theorem) it has maximum and minimum; let's denote them by M and $m \implies$

$$(b-a)\,m \leq \int_a^b f(x)\,dx \leq (b-a)\,M$$

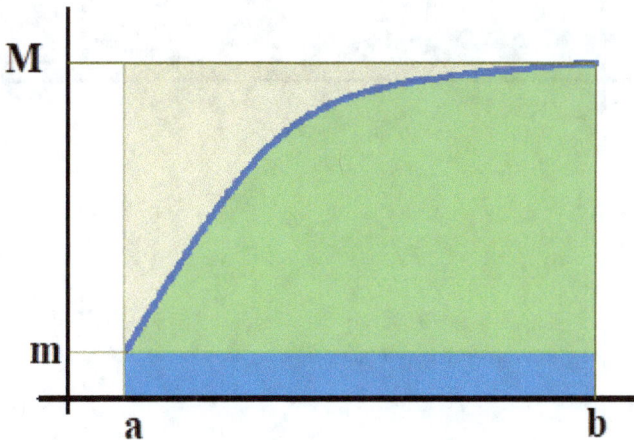

$\implies$ (intermediate values theorem) $f(x)$ gets all the values between m and $M \implies$

$$\exists c \in \,]a;b[\,: \int_a^b f(x)\,dx = (b-a)\,f(c)$$

■

Geometric interpretation:

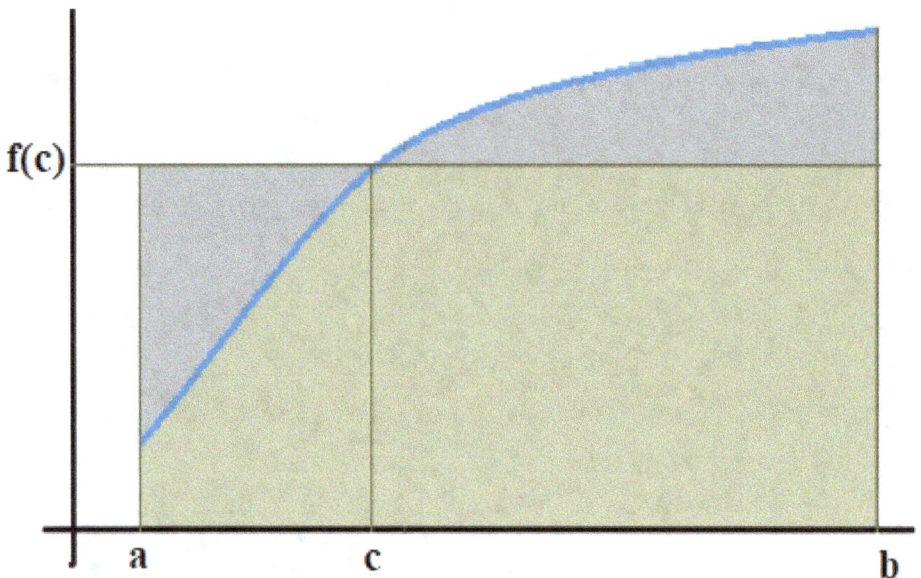

$\int_a^b f(x)\,dx$ is the area underlying the graph of the function and overlying the horizontal axis, while $(b-a)f(c)$ is the area of the rectangle with sides $[a;b]$ (width) and $f(c)$ (height); if we proprerly choose c the area of the rectangle will be equal to the integral (grey parts are equal). If the function is not continuous the point c may not exist. Consider the following picture:

the integral of the (step) function is the area in green, and is equal to the area of the red rectangle having width $(b-a)$ and height h but there are no points c such that $f(c) = h$.

Definition 14.5 *The function*

$$F(x) = \int_a^x f(t)\, dt$$

is said **integral function**; *it is sometime called cumulative function.*

Definition 14.6 *any function $G(x) : G'(x) = f(x)$ is called indefinite integral or primitive (sometimes antiderivative) of $f(x)$; the indefinite integral is denoted by $\int f(x)\, dx$.*

Theorem 14.2 *If two functions have the same derivative everywhere then they differ by an additive constant.*

$$[F'(x) = G'(x) \forall x] \implies [F'(x) - G'(x) = 0] \implies [D(F-G)(x) = 0] \implies [(F-G)(x) = k]$$

Theorem 14.3 *(Fundamental th. of integrals or Torricelli-Barrow th.)*

$$f(x) \text{ continuous in } [a;b] \implies \begin{cases} F(x) \text{ is smooth in }]a;b[\\ F'(x) = f(x) \end{cases}$$

Proof. $\lim_{\Delta x \to 0} \dfrac{F(x+\Delta x) - F(x)}{\Delta x} = \lim_{\Delta x \to 0} \dfrac{\int_a^{x+\Delta x} f(t)\,dt - \int_a^x f(t)\,dt}{\Delta x} =$

$= \lim_{\Delta x \to 0} \dfrac{\int_a^x f(t)\,dt + \int_x^{x+\Delta x} f(t)\,dt - \int_a^x f(t)\,dt}{\Delta x} = \lim_{\Delta x \to 0} \dfrac{\int_x^{x+\Delta x} f(t)\,dt}{\Delta x} = \lim_{\Delta x \to 0} f(c) = f(x)$

the equality

$$\lim_{\Delta x \to 0} \dfrac{\int_x^{x+\Delta x} f(t)\,dt}{\Delta x} = \lim_{\Delta x \to 0} f(c)$$

holds because of the mean value theorem (the function f is continuous) and since $x < c < x + \Delta x$ when $\Delta x \longrightarrow 0$ $c \longrightarrow x$. ∎

Corollary 14.1 *(main consequence of the fundamental theorem)*

If $f(x)$ is continuous in $[a;b] \implies \int_a^b f(x)\,dx = F(b) - F(a)$

where $F(x)$ is any primitive (indefinite integral) of $f(x)$ that is $F'(x) = f(x)$.

Example 14.1 *compute* $\int_0^1 2x\,dx$

The reader can easily realize that the integral is the area of the right triangle represented in the picture below:

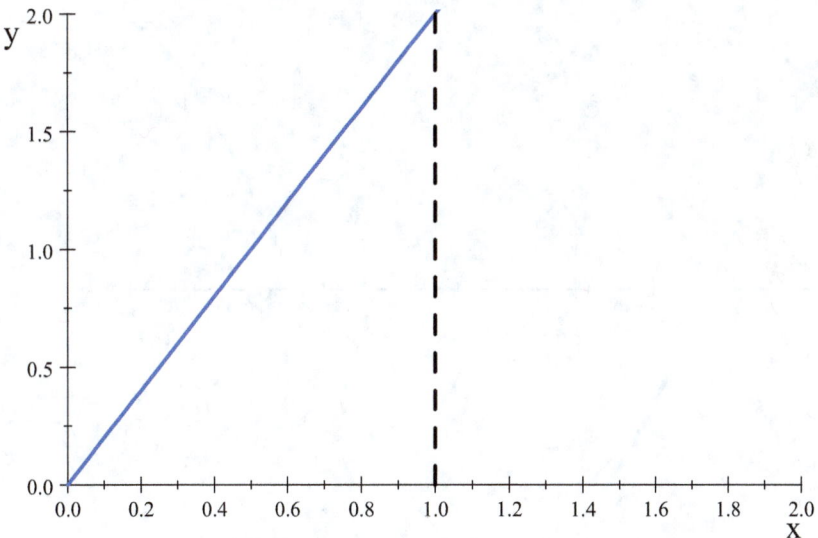

It is possibile to compute this area using our knowledge of elementary geometry; hence:

$$\int_0^1 2x\,dx = \frac{b \cdot h}{2} = \frac{1 \cdot 2}{2} = 1$$

We can compute the same integral using the fundamental theorem (it is not hard to understand that $2x$ is the derivative of the function $y = x^2 + c$):

$$\int_0^1 2x\,dx = \left[x^2\right]_0^1 = 1^1 - 0^2 = 1$$

Example 14.2 *compute* $\int_0^{\frac{\pi}{2}} \cos x\,dx$

$$\int_0^{\frac{\pi}{2}} \cos x\,dx = [\sin x]_0^{\frac{\pi}{2}} = \sin\frac{\pi}{2} - \sin 0 = 1$$

The area we have just computed is the yellow are in the picture below:

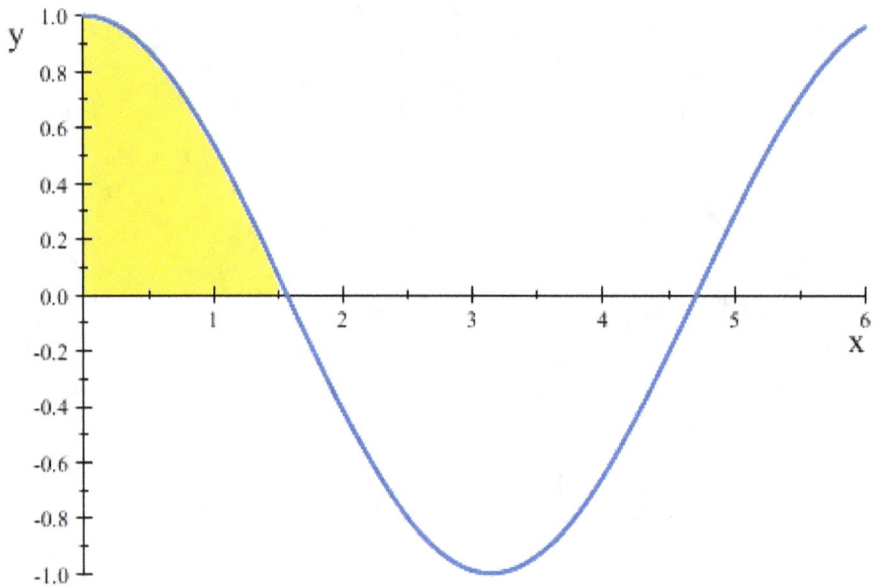

14.4 Table of "immediate" integrals

When computing a definite integral the first step (and quite often the hardest one) is to determine a primitive function. We refer to this step as the solution of an indifinite integral; since all the primitives of a certain function differ by an additive constant an indefinite integral has infinite many solutions. For this reason we write:

$$\int f(x)\,dx = F(x) + c$$

c is a constant wich may assume infinite many values.

Here we present a list of immediate integrals (the list follows from the derivative of elementary functions):

$$\int k dx = kx + c$$

$$\int x^\alpha dx = \frac{x^{\alpha+1}}{\alpha+1} + c \qquad (\alpha \neq -1)$$

$$\int x^{-1} dx = \int \frac{1}{x} dx = \log|x| + c$$

About last integral: I kindly ask the reader to compute the derivatives of the functions $y = \log x$ and $y = \log(-x)$.

$$\int e^x dx = e^x + c$$

$$\int a^x dx = \frac{a^x}{\log a} + c$$

$$\int \sin x dx = -\cos x + c$$

$$\int \cos x dx = \sin x + c$$

$$\int \frac{1}{1+x^2} dx = \arctan x + c$$

$$\int \frac{1}{\sqrt{1-x^2}} dx = \arcsin x + c$$

$$\int \frac{1}{\cos^2 x} dx = \int 1 + \tan^2 x = \tan x + c$$

If the reader is not confident with integral calculus I *strongly strongly* recommend to verify the rules by computing the derivatives of the proposed solutions.

14.4.1 "Almost immediate" integrals

The rules I'm going to present come from the rules presented in the previous section and the chain rule.

$$\int [f(x)]^\alpha \cdot f'(x) = \frac{[f(x)]^{\alpha+1}}{\alpha+1} + c \qquad (\alpha \neq -1)$$

$$\int \frac{f'(x)}{f(x)} dx = \log|f(x)| + c$$

$$\int e^{f(x)} f'(x) dx = e^{f(x)} + c$$

$$\int a^{f(x)} \cdot f'(x) dx = \frac{a^{f(x)}}{\log a} + c$$

$$\int \sin[f(x)] \cdot f'(x) dx = -\cos[f(x)] + c$$

$$\int \cos[f(x)] \cdot f'(x) dx = \sin[f(x)] + c$$

$$\int \frac{f'(x)}{1 + [f(x)]^2} dx = \arctan[f(x)] + c$$

$$\int \frac{f'(x)}{\sqrt{1 - [f(x)]^2}} dx = \arcsin[f(x)] + c$$

Example 14.3

$$\int e^{\sin x} \cos x \, dx = e^{\sin x} + c$$

$$\int \frac{1}{x+1} dx = \log|x+1| + c$$

$$\int -\tan x \, dx = \int \frac{-\sin x}{\cos x} dx = \log|\cos x| + c$$

14.5 Integration techniques

There are several integration techniques; when computing integral very often the hardest thing is to choose the right method to apply. Think as an example to:

$$\int \frac{e^x}{1 + e^x} dx \quad \text{and} \quad \int \frac{e^x}{1 + e^{2x}} dx$$

they seems quite similar, but solutions are very different:

$$\log(1 + e^x) + c \quad \text{and} \quad \arctan e^x + c$$

Please derive the primitive if you are not sure about what you have just read.

Only experience may help the poor student who starts studying this topic, and the best way to build experience is to practise (in most cases practise is better than a good teacher). Don't forget!

14.5.1 Applying the rules of the derivatives

Since integral computation is strongly related to derivatives it is very important to keep in mind two particular rules which can be traslated from derivatives to integrals:

$$D[k \cdot f(x)] = k \cdot D[f(x)]$$
$$\int k \cdot f(x)\, dx = k \cdot \int f(x)\, dx$$

and

$$D[f(x) + g(x)] = D[f(x)] + D[g(x)]$$
$$\int [f(x) + g(x)]\, dx = \int [f(x)]\, dx + \int [g(x)]\, dx$$

Example 14.4 *considering the first rule:*

$$\int 5x^3\, dx = 5\int x^3\, dx = 5\frac{x^4}{4} + c$$

$$\int \frac{5}{\sqrt[3]{x}}\, dx = 5\int \frac{1}{\sqrt[3]{x}}\, dx = 5\int x^{-\frac{1}{3}}\, dx = 5\frac{x^{\frac{2}{3}}}{\frac{2}{3}} + c$$

Considering the second one:

$$\int \frac{x^2 + 2x + 3}{x}\, dx = \int x\, dx + \int 2\, dx + 3\int \frac{1}{x}\, dx = \frac{x^2}{2} + 2x + 3\log|x| + c$$

$$\int \tan^2 x\, dx = \int \left[1 + \tan^2 x - 1\right]\, dx = \tan x - x + c$$

14.5.2 Integration by parts

$$\boxed{\int f(x) \cdot g'(x)\, dx = f(x) \cdot g(x) - \int f'(x) \cdot g(x)\, dx}$$

The Proof is based on the rule of derivation of a product:

$$D[f \cdot g] = f' \cdot g + f \cdot g'$$

$$\int D[f \cdot g] = \int [f' \cdot g + f \cdot g']$$

$$f \cdot g = \int f' \cdot g + \int f \cdot g'$$

$$f \cdot g - \int f' \cdot g = \int f \cdot g'$$

Example 14.5 *let's consider*

$$\int xe^x\, dx$$

Let's set $f(x) = x$ and $g'(x) = e^x$; it follows $f'(x) = 1$ and $g(x) = e^x$ then

$$\int xe^x\, dx = xe^x - \int e^x\, dx = xe^x - e^x + c$$

$$\int \log x\, dx$$

Now let's set $f(x) = \log x$ and $g'(x) = 1$; it follows $f'(x) = \frac{1}{x}$ and $g(x) = x$ then

$$\int \log x\, dx = x \log x - \int \not{x} \cdot \frac{1}{\not{x}}\, dx = x \log x - x + c$$

Integration by substitution

Sometimes it may be difficult to solve an integral using the rules presented up to now; integration by substitution is based on the chain rule and helps in writing an integral in a simpler way.

$$\boxed{\int f(g(x)) \cdot g'(x)\, dx = \int f(t)\, dt}$$

where $t = g(x)$ and $dt = g'(x)\, dx$.

Example 14.6

$$\int \frac{\cos x}{1+(\sin x)^2} dx = \int \frac{1}{1+t^2} dt = \arctan t + c = \arctan(\sin x) + c$$

14.6 Improper integrals

The integrals

$$\int_a^{+\infty} f(x)\, dx$$

$$\int_{-\infty}^b f(x)\, dx$$

are called improper integrals; if:

$$\lim_{b \to +\infty} \int_a^b f(x)\, dx \text{ exists and is finite}$$

or

$$\lim_{a \to -\infty} \int_a^b f(x)\, dx \text{ exists and is finite}$$

we say that the improper integral exists.

If the limits do not exist or are not finite the integrals do not exist.

Example 14.7

$$\int_0^{+\infty} e^{-x} dx = \lim_{b \to +\infty} \int_0^b e^{-x} dx = \lim_{b \to +\infty} \left[-e^{-x}\right]_0^b =$$

$$= \lim_{b \to +\infty} \left(e^0 - e^{-b}\right) = 1$$

and so the improper integral exists.

$$\int_1^{+\infty} \frac{1}{1+x^2} dx = \lim_{b \to +\infty} \int_1^b \frac{1}{1+x^2} dx = \lim_{b \to +\infty} [\arctan x]_1^b =$$
$$= \lim_{b \to +\infty} (\arctan b - \arctan 1) = \frac{\pi}{4}$$

and so the improper integral exists.

$$\int_1^{+\infty} \frac{1}{x} dx = \lim_{b \to +\infty} \int_1^b \frac{1}{x} dx = \lim_{b \to +\infty} [\log x]_1^b =$$
$$= \lim_{b \to +\infty} (\log b - \log 1) = +\infty$$

hence the improper integral does not exist.

Chapter 15

Functions of several variables

The functions analyzed up to now were depending on a single variable; in this part we are going to face functions depending on more than one variable. These functions are widely used in economics: think as an example to the function describing the quantity of a good produced by a company; the output depends on several inputs, the output is the unique depending variable and the inputs are the independent variables (raw materials, capital, labor and so on).

15.1 Domain

The symbolic statement of a function depending on n independent variables may be:

$$y = f(x_1, x_2, ..., x_n)$$

Since the sequence $(x_1, x_2, ..., x_n)$ is a point of $\mathbb{R}^n$ we'll use the notation

$$f : \mathbb{R}^n \longrightarrow \mathbb{R}$$

Now the domain of such a function is a subset of $\mathbb{R}^n$ while the graph is a subset of $\mathbb{R}^{n+1}$.

Here we will:

1. plot the domain of the functions $\mathbb{R}^2 \longrightarrow \mathbb{R}$;

2. define partial derivatives;

3. compute partial derivatives;

4. determine the extrema of functions $\mathbb{R}^2 \longrightarrow \mathbb{R}$.

15.2 The graph of functions depending on several variables

It is not possible to represent graphically the graph of a function $\mathbb{R}^n \longrightarrow \mathbb{R}$ on a sheet of paper when $n \geq 2$; in the special case where $n = 2$ we can somehow represent a three-dimensional surface on a paper (a two dimensional space).

As an example let's consider the function $f(x, y) = x^2 + y^2$; its graph is something similar to:

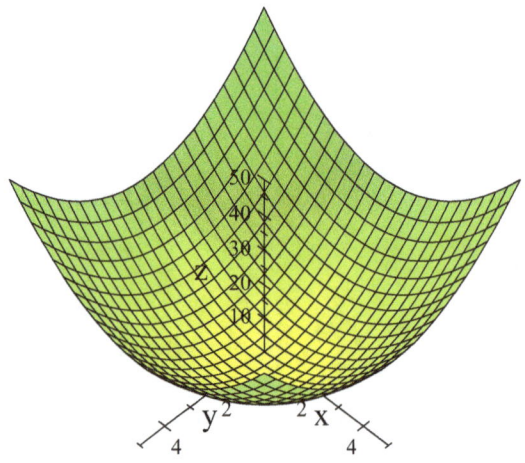

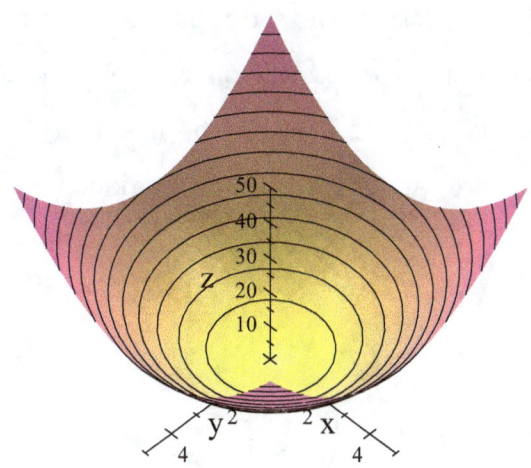

The function has been plotted twice to help the reader in viewing the graph.

Obviously it is not possible to give a representation of a function $f: \mathbb{R}^n \longrightarrow \mathbb{R}$ on a sheet of paper when $n > 2$. To realize this imagine a function $f: \mathbb{R}^3 \longrightarrow \mathbb{R}$; it's graph is a surface in the 4−mensional space! Mathematicians are able to do many things in the 4−dimensional space, but neither them are able to graphically represent a function in this space.

15.2.1 Plotting the domain of a function $\mathbb{R}^2 \longrightarrow \mathbb{R}$

Coming back to the case where $n = 2$, the domain is a subset of $\mathbb{R}^2$, we saw in the previous section that it is possible to reduce its 3-dimensional graph in the 2-dimensional space; since the domain is a subset of $\mathbb{R}^2$ it is possible to represent it.

Let's do it considering the function $f(x, y) = \sqrt{y - x^2 + 1} + \sqrt{x - y - 1}$.

The first root exists only when $y - x^2 + 1 \geq 0$ that is when $y \geq x^2 - 1$. This parabola is represented in the picture; the set of points satisfying such inequality is the upper graph of the parabola; the second square root exists if $y \leq x - 1$; the line in the picture has equation $y = x - 1$ and the inequality is satisfied by its lower graph. Both inequalities are satisfied in the blue region.

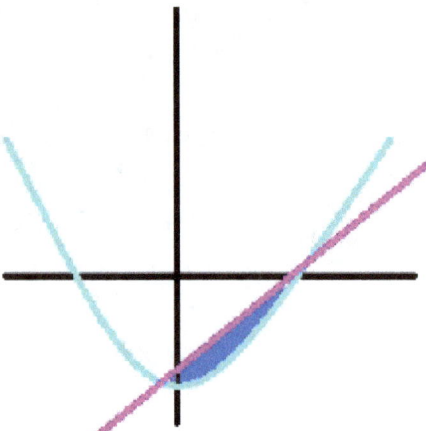

Proposition 1 we can apply all the rules we already know about the functions depending on a single variable

considering the other variables as constants.

15.3 Partial derivatives

While considering a function $\mathbb{R} \longrightarrow \mathbb{R}$ we saw that its derivative is a limit:
$$f'(c) = \lim_{x \to c} \frac{f(x) - f(c)}{x - c}$$
and since $c \in \mathbb{R}$ it was possible to approch the point moving in only one direction (the x approaches c moving on the real line either from the left or from the right).

Now things are much different **c** is a point belonging to $\mathbb{R}^n$ and we can approach the point from infinite many directions; for this reason we can define (and compute) infinite many derivatives in a point.

Here we consider only the derivatives we can get approching the point **c** moving along directions that are parallel to the cartesian axes. In this way for a function $f : \mathbb{R}^n \longrightarrow \mathbb{R}$ **y** $= f(x_1, x_2, ..., x_n)$ we can compute n derivatives: the derivatives with respect to x_i, $i \in \{1, 2, ..., n\}$; these derivatives are called partial derivatives and the notation used are f'_{x_i} or $\frac{\partial f}{\partial x_i}$ (read: derivative of f with respect to variable x_i).

In order to better understand this concept let's think to a function $f : \mathbb{R}^2 \longrightarrow \mathbb{R}$; in this case we have a dependent variable and two inputs: $z = f(x, y)$; now we can derive f either with respect to x (moving along a direction parallel to the x axis) or with respect to y (moving along a direction parallel to the y axis)

Definition 15.1 *(Partial derivative) let* $\mathbf{x} \in \mathbb{R}^n$ *then*

$$f'_{x_i}(\mathbf{x}) = \frac{\partial f}{\partial x_i} \stackrel{def}{=} \lim_{\Delta x_i \to 0} \frac{f(x_1, ..., x_i + \Delta x_i, ..., x_n) - f(x_1, ..., x_i, ..., x_n)}{\Delta x_i}$$

if this limit does exist and is finite. Often we'll use the simplified notation f'_{x_i}.

Proposition 15.1 *we can compute partial derivatives applying all the rules we already know about the functions depending on a single variable considering the other variables as constants.*

Example 15.1 *Let's consider:*

$$f(x, y) = 3x^5 y + xy^2 + 2x - 5y$$

then:

$$f'_x(x, y) = 3 \cdot 5x^4 \cdot y + 1 \cdot y^2 + 2 - 0$$

As the reader can realize variable y is treated as a constant: it remains unchanged when it is multiplying some functions of x (as in the first and the second term) and its derivative is 0 when it is just an additive term (last one).

$$f'_y(x, y) = 3x^5 \cdot 1 + x \cdot 2y + 0 - 5$$

Definition 15.2 *The symbol ∇f is the gradient of f that is the vector of the partial derivatives:*

$$\nabla f \stackrel{def}{=} \begin{pmatrix} f'_{x_1} \\ f'_{x_2} \\ \dots \\ f'_{x_n} \end{pmatrix}$$

Example 15.2 *The gradient of $f(x,y,z) = x^2 + y^2 + x^3 y^3 + z^x$ is*

$$\nabla f = \begin{pmatrix} f'_x \\ f'_y \\ f'_z \end{pmatrix} = \begin{pmatrix} 2x + 3x^2 y^3 + z^x \log z \\ 2y + 3x^3 y^2 \\ xz^{x-1} \end{pmatrix}$$

15.4 Extrema of vectorial functions

The procedure to determine the extrema of a vectorial function is somehow based on the same idea underlying the search of the extrema for functions $\mathbb{R} \to \mathbb{R}$:

Let $f : \mathbb{R} \to \mathbb{R}$ be a continuous and smooth function. There are sufficient conditions to determine its extrema. These conditions are called first order condition (involving the derivative of the first order) and second order condition (involving the derivative of the second order).

The first order condition is satisfyied if the derivative is equal to zero; let's recall that if $f'(c) = 0$ then c is called stationary point and c may be a minimum, a maximum or a turning point with horizontal tangent line.

If c is a stationary point; the second order conditions are:

if $f''(c) > 0$ then c is a minimum point

if $f''(c) < 0$ then c is a maximum point

if $f''(c) = 0$ then we can conclude nothing and further analysis are necessary: c may be a maximum, a minimum or a turning (inflection) point with horizontal tangent line.

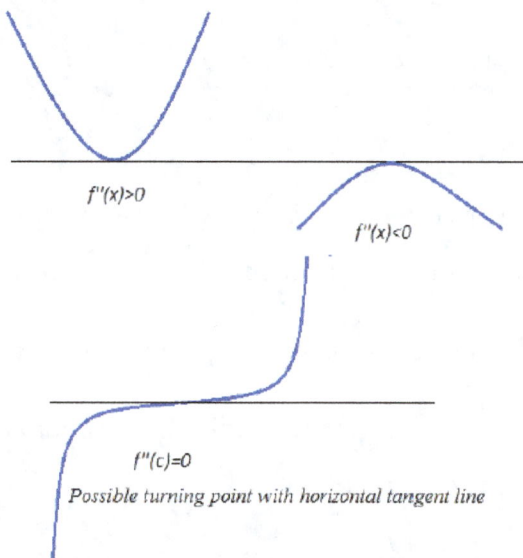

Possible turning point with horizontal tangent line

First order condition implies that the tangent line is horizontal; if the derative of the second order is positive the function is convex and it follows that the point is a minimum (see above picture); if the derative of the second order is negative the function is concave and the point is a maximum.

If f is a function $\mathbb{R}^n \longrightarrow \mathbb{R}$ like $y = f(x_1; x_2)$ in the case that $n = 2$, the first order conditions imply the existence of an horizontal tangent plane (or hyperplane):

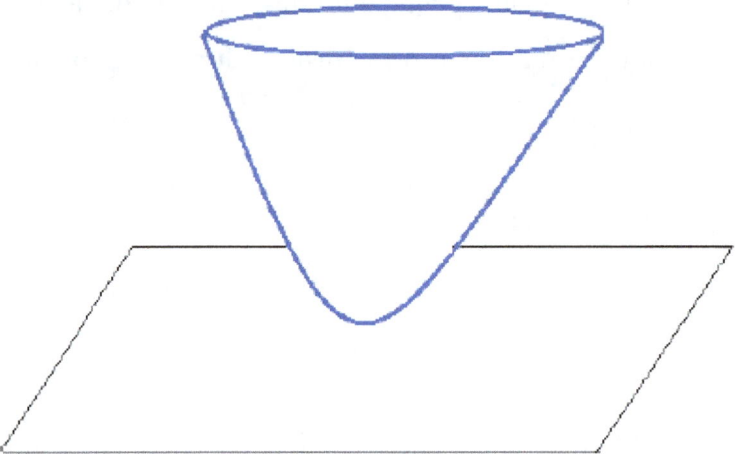

The first order conditions require that all the partial derivatives must be equal to zero (in other words the gradient must be equal to the null vector).

Here is the full analysis for functions $f : \mathbb{R}^2 \longrightarrow \mathbb{R}$, $z = f(x;y)$;

First order conditions (necessary conditions for a maximum or minimum point laying in the interior of the domain)

$$\nabla f = \begin{pmatrix} f'_x \\ f'_y \end{pmatrix} = \begin{pmatrix} 0 \\ 0 \end{pmatrix}$$

that is all the partial derivatives must be equal to zero. A point satisfying the first order conditions is called stationary point.

Before dealing with the second order conditions we need some more tools. First of all we have to refer ot the Schwartz theorem.

Theorem 15.1 *(Schwartz theorem): given a function $f : \mathbb{R}^n \longrightarrow \mathbb{R}$, if its second partial derivatives f''_{xy} and f''_{yx} do exist in a neighborhood of $\mathbf{x}$ and they are continuous in the point then they are equal i.e.*

$$f''_{xy}(\mathbf{x}) = f''_{yx}(\mathbf{x})$$

Hessian matrix:

The Hessian matrix is a square matrix containing the second partial derivatives; in the case of a function $f : \mathbb{R}^2 \longrightarrow \mathbb{R}$ we have the following 2×2 matrix:

$$\mathbf{H} = \begin{bmatrix} f''_{xx} & f''_{xy} \\ f''_{yx} & f''_{yy} \end{bmatrix}$$

We use $\mathbf{H}$ to denote the matrix containing the symbolic derivatives, $\mathbf{H}(x_0; y_0)$ to denote the matrix containing the derivatives computed in the point $(x_0; y_0)$. To better understand these concepts let's consider the following:

Example 15.3 *Let's consider $f(x, y) = x^4 + y^3$ and $P = (1; 2)$; then:*

$$\nabla f = \begin{pmatrix} 4x^3 \\ 3y^2 \end{pmatrix}$$

$$\mathbf{H} = \begin{bmatrix} 12x^2 & 0 \\ 0 & 6y \end{bmatrix}$$

$$\mathbf{H}(P) = \begin{bmatrix} 12 & 0 \\ 0 & 12 \end{bmatrix}$$

We now define the determinant $(\det(\mathbf{H}))$ and the trace $(tr(\mathbf{H}))$ of the hessian matrix:

$$\det(\mathbf{H}) = f''_{xx} \cdot f''_{yy} - f''_{yx} \cdot f''_{xy}$$
$$tr(\mathbf{H}) = f''_{xx} + f''_{yy}$$

Second order conditions (sufficient conditions to have a minimum or maximum point):

Let $(x_0; y_0)$ be a stationary point;

if: $det(\mathbf{H}) > 0$ and $tr(\mathbf{H}) > 0$ then $(x_0; y_0)$ is a minimum point;

if: $det(\mathbf{H}) > 0$ and $tr(\mathbf{H}) < 0$ then $(x_0; y_0)$ is a maximum point;

if: $det(\mathbf{H}) < 0$ $(x_0; y_0)$ is neither a maximum nor a minimum (it is a saddle point);

if: $det(\mathbf{H}) = 0$ and $tr(\mathbf{H}) < 0$ $(x_0; y_0)$ isn't a minimum;

if: $det(\mathbf{H}) = 0$ and $tr(\mathbf{H}) > 0$ $(x_0; y_0)$ isn't a maximum.

Most of the authors uses a different, but equivalent, statement for the second order conditions:

Let $(x_0; y_0)$ be a stationary point;

if: $det(\mathbf{H}) > 0$ and $f''_{xx} > 0$ then $(x_0; y_0)$ is a minimum point;

if: $det(\mathbf{H}) > 0$ and $f''_{xx} < 0$ then $(x_0; y_0)$ is a maximum point;

if: $det(\mathbf{H}) < 0$ $(x_0; y_0)$ is neither a maximum nor a minimum (it is a saddle point).

Exercise 15.1 *Let $f : \mathbb{R}^2 \to \mathbb{R}$; proove that if its second partial derivatives exist in a neighborhood of $\mathbf{x}$ and they are continuous in $\mathbf{x}$ then it is not possible to have in $\mathbf{x}$ $det(\mathbf{H}) > 0$ and $tr(\mathbf{H}) = 0$.*

Example 15.4 $f(x; y) = xy$ *has only one stationary point:* $\begin{pmatrix} x \\ y \end{pmatrix} = \begin{pmatrix} 0 \\ 0 \end{pmatrix}$ *(please verify);* $H(0,0) = \begin{bmatrix} 0 & 1 \\ 1 & 0 \end{bmatrix}$; $\det(H) = -1 < 0$ *and so the origin of the cartesian plane is a saddle point.*

Example 15.5 *A graphical representation of a saddle point:*

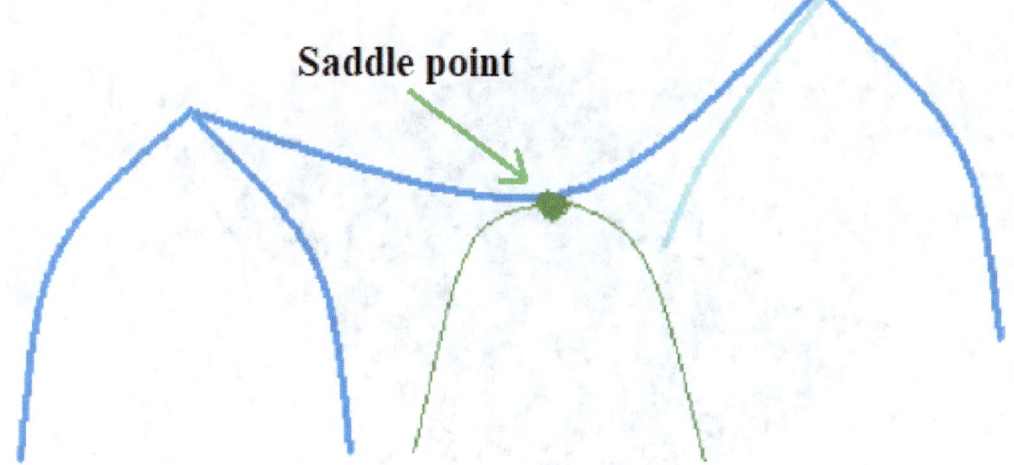

Saddle point

Example 15.6 *Determine the extrema of the function $f(x;y) = x^2 + y^2 - 2x + 6y$.*

The first order conditions are

$$f'_x = 2x - 2 = 0 \iff x = 1$$
$$f'_y = 2y + 6 = 0 \iff y = -3$$

The unique stationary point is: $(1;-3)$;

the hessian matrix is $H(1;-3) = \left[\begin{pmatrix} 2 & 0 \\ 0 & 2 \end{pmatrix}\right]$ and we have:

$det(H) = 4 > 0$ $tr(H) = 4 > 0$

hence $(1;-3)$ is a minimum point.

Graph of $f(x;y) = x^2 + y^2 - 2x + 6y$

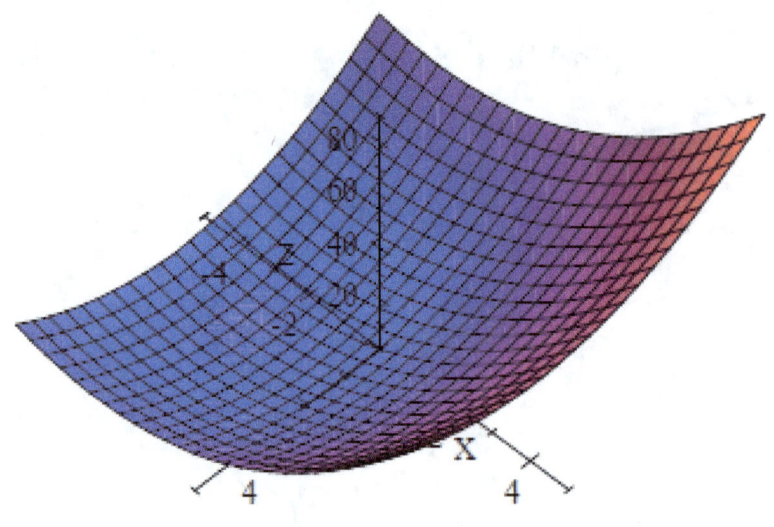

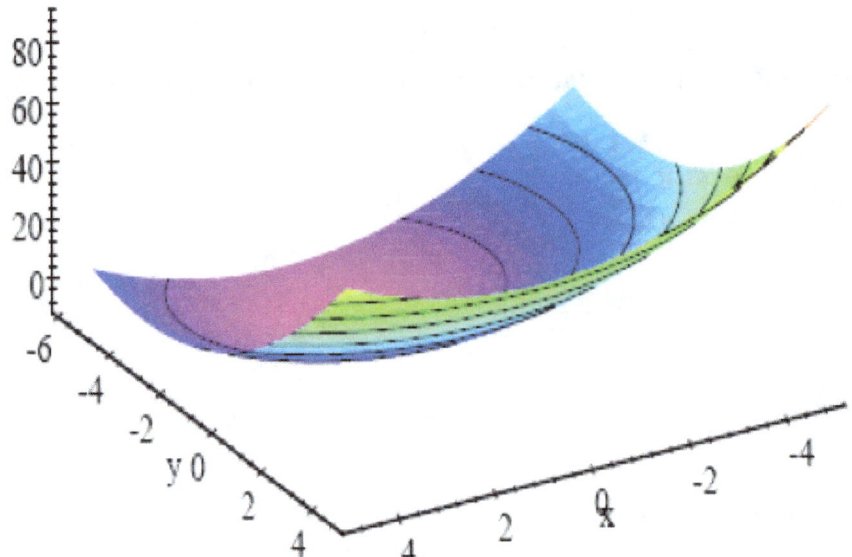

Example 15.7 *Let's consider $f(x;y) = -x^2 - y^2$ and let's determine its extrema; the first order conditions are:*

$$f'_x = -2x = 0 \iff x = 0$$
$$fy = -2y = 0 \iff y = 0$$

$\begin{pmatrix} 0 \\ 0 \end{pmatrix}$ is the unique stationary point; $\mathbf{H} = \begin{bmatrix} -2 & 0 \\ 0 & -2 \end{bmatrix}$

$det(\mathbf{H}) = 4 > 0$, $tr(\mathbf{H}) = -4 < 0$ hence $\begin{pmatrix} 0 \\ 0 \end{pmatrix}$ is a maximum point.

Example 15.8 *Here is the graph of $f(x;y) = -x^2 - y^2$*

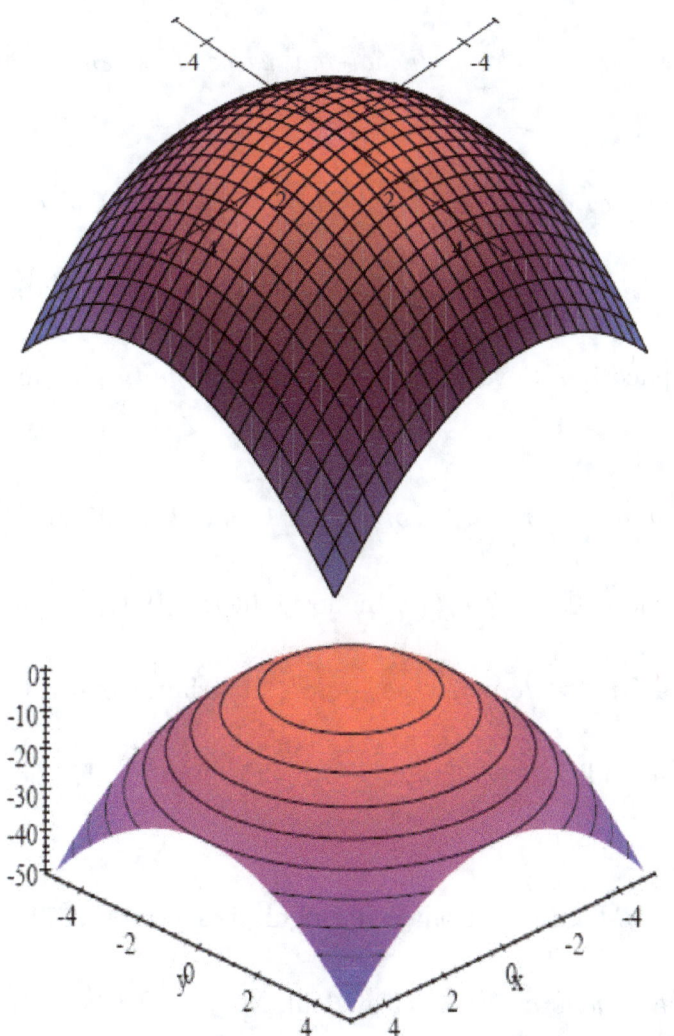

Bibliography

[1] Barozzi G. C. e Corradi C., "*Corso di matematica per le scienze economiche*", il Mulino, Bologna 1988.

[2] Calzone O., "*Matematica di base partendo da zero*", Ottavio Calzone (2020).

[3] Calzone O., "*Capire l'algebra in 29 giorni*", Ottavio Calzone (2020).

[4] Conti F., Acquistapace P. e Savojini A., "*Analisi matematica*", Mc Graw-Hill, Milano (2001).

[5] Malafarina G., "*Matematica per i precorsi*", Mc Graw-Hill, Milano (2003).

[6] Marcellini P. e Sbordone C., "*Calcolo*", Liguori editore (1992).

[7] Marini C. e Scianna G., "*Matematica generale*", libreriauniversitaria.it (2013).

[8] Privileggi F., "Compendio di *Matematica per l'Economia*", Edizioni Simone, Arzano (2007).

[9] Prodi G., "*Analisi matematica*", Bollati Boringhieri, Torino (1987).

[10] Ricci G., "*Matematica generale*", Mc Graw-Hill, Milano (2001).

[11] Zezza P. "*Metodi matematici per le scienze economiche e aziendali*", Carocci editore, Roma (2009).

[12] Zwirner G., "*Esercizi di nalisi matematica parte seconda*", Cedam, Padova (1977).

Suggested websites

- https://docenti-deps.unisi.it/marcolonzi/didattica/matematica-generale/ (in Italian).

- https://docenti-deps.unisi.it/samuelericcarelli/matematica-generale/ (in Italian).

- https://matematicainpalio.blogspot.com/ (in Italian).

- https://www.youmath.it/esercizi.html (in Italian).

www.ingramcontent.com/pod-product-compliance
Lightning Source LLC
Chambersburg PA
CBHW082104220526
45472CB00009B/2044